Surfing the Global Tide

Surfing the Global Tide

Automotive Giants and How to Survive Them

Michael S. Wynn-Williams

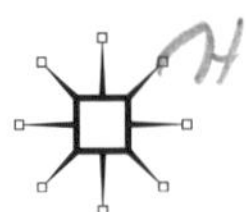

First published 2009 by
PALGRAVE MACMILLAN

Palgrave Macmillan in the UK is an imprint of Macmillan Publishers Limited, registered in England, company number 785998, of Houndmills, Basingstoke, Hampshire RG21 6XS.

Palgrave Macmillan in the US is a division of St Martin's Press LLC, 175 Fifth Avenue, New York, NY 10010.

Palgrave Macmillan is the global academic imprint of the above companies and has companies and representatives throughout the world.

Palgrave® and Macmillan® are registered trademarks in the United States, the United Kingdom, Europe and other countries

ISBN-13: 978–0–230–57924–8
ISBN-10: 0–230–57924–8

This book is printed on paper suitable for recycling and made from fully managed and sustained forest sources. Logging, pulping and manufacturing processes are expected to conform to the environmental regulations of the country of origin.

A catalogue record for this book is available from the British Library.

A catalog record for this book is available from the Library of Congress.

10 9 8 7 6 5 4 3 2 1
18 17 16 15 14 13 12 11 10 09

Printed and bound in Great Britain by
CPI Antony Rowe, Chippenham and Eastbourne

Contents

List of Tables

List of Figures

List of Abbreviations

AFC	average fixed costs
ATM	automatic transfer machine
BAe	British Aerospace
BIW	body-in-white process
BL	British Leyland
BLMC	British Leyland Motor Corporation
BMC	British Motor Corporation
BMH	British Motor Holdings
CAD	computer aided design
CKD	complete knock down
CVCC	compound vortex controlled combustion
DMNC	diversified multinational corporation
DTI	Department of Trade and Industry
FDI	foreign direct investment
GM	General Motors
GMAC	General Motors Acceptance Corporation
GRP	glass reinforced plastic
HO	Heckscher-Ohlin
HUM	Honda UK Manufacturing
IPR	intellectual property rights
IR	industrial relations
IT	information technology
IVJV	international vertical joint venture
JAMA	Japan Automobile Manufacturers Association
JIT	just-in-time
JV	joint venture
LAC	long-run average cost
MC	marginal costs
MEPS	minimum efficient plant size
MES	minimum efficient scale
MES_p	prototypical minimum efficient scale
MIC/MIS	minimum inventory control/strategy
MNE	multinational enterprise

MPV	multipurpose vehicle
MR	marginal revenue
NAC	Nanjing Automobile Corporation
NAFTA	North American Free Trade Agreement
NEB	National Enterprise Board
OICA	International Organization of Motor Vehicle Manufacturers
OLI	ownership, location, and internalization
PAG	Premier Automotive Group
PVH	Phoenix Venture Holdings
RMS	reconfigurable manufacturing system
SAIC	Shanghai Automotive Industry Corporation
SATC	short-run, average, total cost
SAVC	short-run, average, variable cost
SMC	short-run marginal cost
SUV	sports utility vehicle
TCA	transaction cost analysis
TGWU	Transport and General Workers Union
TPS	Toyota Production System
TQ	total quality
VJV	vertical joint venture
WTO	World Trade Organization

ACKNOWLEDGEMENTS

For the first three years of this research program I was fortunate to gain funding from the Engineering and Physical Sciences Research Council (EPSRC), which included several trips to Japan. I was later sponsored by Trend Tracker Ltd and subsequently IHS Global Insight allowed me to use some of their historical data. Other crucial sources of data were kindly provided by the Automotive News Data Centre and the global industry organization, the International Organization of Motor Vehicle Manufacturers (OICA).

In the course of the research I would like to thank everyone who allowed me to interrogate them. All interviewees gave their time freely and I have nothing but fond memories of the fieldwork. At Cardiff Business School, Dr Paul Nieuwenhuis assisted with the historical material and can be said to be the originator of the Budd Paradigm that inspired this research. Professor Garel Rhys was a beacon of knowledge and had a far greater impact on the research than the brevity of his advice might suggest. Most of all, I am grateful to Professor Trevor Boyns without whose tireless guidance this book would never have attained the standard that it did.

Furthermore, throughout this six-year sentence of penal servitude in the cause of automotive research, I have been sustained by a host of people, many of whom will not be aware of the support they have given me. Somewhere in the background there was always Roger Waters, while Jyoti Butel and Stefan Lang provided me with a personal lift when I needed it most. Ultimately, though, there is one who has stuck by me over the past decade with a project of her own: me. Well, Kerry Ann Kludsikofsky, your work will never end.

MICHAEL WYNN-WILLIAMS

CHAPTER 1

Raising the leviathan

Automobile manufacturing has grown into an industry so monstrous that it defines the twentieth century like no other human endeavor. It has shaped our society to the extent that it is now difficult to envisage what form the physical landscape would have taken without this network of highways that scar the scenery, linking disparate residential communities with places of work, recreation, and retail. The dependence on the automobile is now so great that if it were to disappear it would reduce many communities to crisis point as they became isolated from all that sustains them.

The industrial network that underpins the automobile has also spread its tentacles throughout the commercial world. The production flow system that characterizes automobile manufacturing was not new even when Henry Ford first implemented it, and neither has Toyota's so-called lean production approach taken it beyond its original principles; but it has acted as an icon for economic efficiency with ramifications far beyond its industrial foundation. Production flows can now be observed even in industries where it was once thought to be impossible, such as shipbuilding and aircraft manufacturing, while lean production has become a totem of efficient activity.

Yet despite the automobile industry's position at the pinnacle of industrial and economic development it is still not well understood. Each new innovation seems to represent a blind step into the unknown as the industry explores an unknowable future. This suggests a chaotic progression as companies fumble in the dark for a route to prosperity, some inevitably losing their way and staggering into the Grimpen Mire of history. At the same time, other companies seem adept at finding a way forward, and it is salutary to note that some of the oldest names in the industry are still competing as fiercely as ever. Nevertheless, the inexorable rise of a new industrial colossus in the shape of Toyota has inspired a fear in the industry that the Japanese have discovered the one true course to the future.

To discover what this route might be, this research attempts to clarify the basic principles upon which the automobile industry is founded by revealing the fundamental structuring forces. Since these are essentially the same for all mass market automobile manufacturers it is possible to arrive at a self-contained and systematic view of the industry, expressed as a paradigm similar to those that structure schools of thought in science and philosophy. The principles embodied in the paradigm then define the limitations of the industry and govern its future path. Using actual data gathered from automotive companies in Japan and the UK, this research has constructed an evolved size and structure for a theoretical automobile company in accordance with the principles of the proposed paradigm. Uniquely, the resultant model quantifies the extent of vertical integration and necessary economies of scale related to each function for a sustainable automobile manufacturer. The model also illustrates the additional advantages that accrue to manufacturers that exceed the prescribed production output quantities.

While this suggests that all automobile companies that fall short of the ideal are doomed to failure, the research also shows how it is possible for suboptimal firms to collaborate in a novel form of partnership: the international vertical joint venture (IVJV). Not only is this of vital intelligence to all automobile firms, but by applying this paradigmatic approach to other industries it is similarly possible to derive an equally systematic understanding of their limitations and future potential. If this is the case, then the automobile industry will remain at the leading edge of industries around the world.

1.1 Growth of a global industry

The world's automotive industry has long played an important economic role. It is now one of the largest industries in the world, but even in 1946 its size prompted Peter Drucker to declare that "the automobile industry stands for industry all over the globe. It is to the twentieth century what the Lancashire cotton mills were to the early nineteenth century: the industry of industries" (Drucker 1972, 176).

Over 30 years later, in 1981, the managing director of Volvo, Pehr G. Gyllenhammar, stated: "I personally believe that the automobile industry marks the limit of the sustainability of industrial society. If you, as a country or a nation, state as a fact that you are not competitive within this industry, then you have also abdicated from industrial society" (Malmberg 1991, 212).

Data from the International Organization of Motor Vehicle Manufacturers (OICA) show that the impact of the industry has not softened since then. In 2007, total vehicle production reached 73.2 million units, an increase of 5.7 percent over 2006, when employment had reached 9 million people in vehicle production and the related component industries. OICA estimates that as much as five times as many more people are employed indirectly, resulting in total employment of around 50 million people. This results in a substantial contribution to the global economy, equivalent to a turnover of €1.9 trillion in 2005, making possible investments in research and development (R&D) and production of around €85 billion. Governments too benefit, with the 26 OICA member countries collecting over €430 billion in revenues.

The greater part of the automotive industry is made up of automobile production, amounting to around 53 million units in 2007 (OICA 2008b). Yet production output is not distributed evenly around the world, with Western Europe taking the larger share at 33 percent of the total. Within the European Union (EU) there has been some shift in production towards the newest member countries so that Romania, Slovenia, and the Czech Republic have gained production while only Germany of the established European countries has maintained some rate of growth (OICA 2006a). Overall, though, total production for the European region has remained fairly steady.

Elsewhere in the world, shifts in general automotive production have been detected. In 2006, automotive vehicle production in North America fell by 2.7 percent while Africa grew by 16.3 percent, although nearly half of these vehicles were trucks. African output was dominated by South Africa which manufactured nearly 340,000 automobiles out of a total of 458,000 for the continent as a whole. The Asia-Pacific region showed growth of 9.3 percent in total automotive output, China being conspicuous with general automotive industry growth of 26.3 percent and automobile output of 4.3 million units. Despite that rate of increase Japan remains the biggest producer in the region with 9.8 million automobiles produced, a rise of 8.2 percent on the year before (JAMA 2007a). Table 1.1 shows the countries that produced over a million units in 2007.

Table 1.1 shows that the expansion of the world's automobile output is rising like a global tide, and one that threatens to overwhelm the weaker industrial centres. It certainly makes depressing reading for followers of the industry in such countries as the US and the UK as they watch their countries sinking ever lower in the rankings. Yet even as emerging economies such as China occupy good positions in the global automobile industry, other countries that have long historical links to the industry continue to be strong players. Germany is still in the top three, while France and the US retain their strategic importance for the industry. In order to grasp the basic factors involved it is necessary, therefore, to specify the unique characteristics of the automobile industry that imbue it with such global importance.

Table 1.1 Top automobile producing countries, 2007

Country	Automobile production output (millions)
Japan	9.9
China	6.4
Germany	5.7
USA	3.9
South Korea	3.7
France	2.6
Brazil	2.4
Spain	2.2
India	1.7
UK	1.5
Canada	1.3
Russia	1.3
Mexico	1.2

Source: OICA (2008b).

1.2 The Budd paradigm lays the foundation

Historically, all vehicles were made using similar methods, an unstressed vehicle body being attached to a load-bearing platform, otherwise known as a frame or chassis. Automobile manufacturing then took on a distinct form of production technology when the all-steel load-bearing unitary body was developed in the early part of the twentieth century. This basic manufacturing process is still the dominant one but it is highly restrictive, both in the form of the final product and in the method of production, so it is best suited to vehicles that have a well defined purpose, i.e. the comfortable and affordable transport of small groups of passengers at their own convenience.

The highly defined state of manufacturing automobiles in this way results in an industry structure that can be described from a systematic viewpoint, known as a paradigm. Paradigms are well known in research fields where they define a particular philosophy, or school of thought. For example, astronomy and astrology might be described as paradigms. While they share an observation of the stars, the fundamental purpose of the two approaches is so distinct that there is little else in common. Where astronomers use radio telescopes to research the structure of stars, astrologers would refer to printed charts that simply tabulate planetary movements. Furthermore, each approach predefines the nature of problems it is designed to tackle: an astrologer could no more hypothesize the existence of a new planet than an astronomer could predict the outcome of a football game.

Indeed, different paradigms are so distinct and self-referential, that when a new approach emerges there is no possibility of the old paradigm evolving into the new one. On the contrary, technological progress tends to be made in fits and starts as the new paradigm subverts the influence of the old one in a revolutionary change. Interestingly, since the new paradigm is based on a distinctly new perspective, and deals with a new set of problems, there is actually scope for the old paradigm to continue until it eventually fades away into irrelevance. Newton's Laws of Motion may be part of an old paradigm but they are still a perfectly acceptable way of moving around the planet. The rival quantum mechanics paradigm answers different questions from a novel perspective and it would require yet another paradigm to resolve the opposing points of view.

Once this paradigmatic approach is accepted it is possible to conceive of their existence in all walks of life: as psychology and psychiatry, for example, or religions like Christianity and Buddhism. In the field of arts it becomes possible to understand why photographers and painters have so little in common, despite the fact that both deal with visual images. Returning this paradigmatic viewpoint to industry, it becomes clear that the Industrial Revolution was only possible once it was discovered how iron could be made available in high volumes. While the original method of fabricating iron, in a blacksmith's forge, was convenient for serving local needs, huge furnaces were ideally suited to large volume output serving equally large-scale factory production. Thus, we find the birth of industrialization and the consolidation of manufacturing on a massive scale.

Similarly, Nieuwenhuis and Wells (1997, 2007) have found that manufacturing automobiles from steel, load-bearing bodies brought a clear definition to the possibilities and limits of automobile production. They have termed this manufacturing process the "Budd Paradigm" after Edward Budd, who, with his partner Joseph Ledwinka, invented the unitary body concept that did away with the separate chassis. It is the associated requirement for a specific production technology that then defines the limits of the paradigm and dictates the ultimate economies of scale for the automobile industry as a whole.

The economies of scale occur in the automotive industry because of the need to use large amounts of capital machinery, a commitment that cannot be avoided due to the physical difficulty in shaping and welding sheet steel. Generally speaking, these metal processing machines have always been able to cope with more production than market demand would justify. This has meant that as global demand has risen, instead of building additional small plants, it has been possible to build ever larger plants housing the same basic production technology. As long as a large plant is working at full capacity, the high costs of operating the plant are spread over the increased output to the extent that each unit can be produced for a lower cost. The availability of economies of scale refers to the way that unit costs fall as larger and larger plants are built. In theory, though, there comes a point when a plant is too big and the problems in running it actually create additional costs; this plant, and any larger ones, are said to be experiencing diseconomies of scale.

In order to exploit fully these economies of scale companies need to invest in large, optimally sized plants. The high risk of doing this from a standing start creates the high entry barriers in the industry that result in the bigger incumbents retaining their place at the top (Altshuler *et al.* 1986). As a consequence, when a new automotive company attempts to secure a place for itself it requires substantial outside assistance, usually from the government. This provides artificial support to elevate indigenous manufacturers to a level where they can compete against the established, global competition. In China the government has done this by obliging foreign manufacturers to enter into joint ventures (JVs) with local firms as a mechanism for transferring technical capability and sharing production output. Chief amongst these Chinese JVs is that between VW and SAIC, which sold 694,406 units in 2006 (*Automotive News* 2007b). Conversely, the smaller domestic Chinese manufacturers that are not in JVs with foreign manufacturers continue to have difficulty simply surviving: they accounted for just 27 percent of the Chinese market in 2006, this being spread over at least 20 different manufacturers (*Business Week* 2007). Even Chery, the largest independent Chinese manufacturer with 7.2 percent of the market, has joint ventures with Fiat and Chrysler along with substantial engineering consultancy input from Lotus of the UK and AVL of Austria (*Automotive News* 2007b; Chery 2007).

There is no sign that these entry barriers to the industry are coming down. On the contrary, new technological challenges for the industry, particularly with regard to safety and exhaust emissions, have meant that product development costs have risen. Bailey gives a range between £400 million and £1 billion (US$800 million and US$2 billion) for bringing a complete new model program to the market: "As a result, large scale production over different brands using a platform-sharing approach is vital to generate the cash for future model development" (Bailey 2007, 139).

The established manufacturers are thought to enjoy advantages in size, in addition to the economies of scale, because they can spread the costs of product development over a greater number of units. This has led to a trend towards higher outputs and consolidation as major manufacturers merge to form larger groups. Maxton and Wormald (1995) predicted that only five global automakers would eventually survive, two in the US and three in Japan. Ford is a case in point, acquiring Jaguar in 1989, Aston Martin in 1994, Volvo in 1999, and Land Rover in 2000 and forming the Premier Automotive Group (PAG) division. There have been similar mergers throughout the industry, such as Daimler and Chrysler in 1998, Renault and Nissan in 1999, and BMW and Rover Group in 1994.

Nevertheless, there have been signs more recently that the industry has overplayed its hand. Since 2000, of those listed above, only the Renault–Nissan relationship has remained intact. PricewaterhouseCoopers (2005) found that by 2004 the value of mergers between vehicle manufacturers had fallen to $2.3 billion (£1.25 billion), which was 9 percent by value of the total merger activity in the automotive related industry and down from 29 percent in 2002. A later annual survey by KPMG (2007) found that while larger firms would continue to hold the advantage, consolidation would no longer be achieved through full bilateral mergers but by networks of strategic alliances. This suggests that firms will be able to access economies of scale jointly while retaining their own organizational structures. *The Economist* (2005) particularly noted that Toyota had come to prominence achieving economies of scale without fundamentally changing its integrated structure.

In 2005, *Automotive News* counted 21 independent automobile manufacturers that were also operating in partnerships and in 2006 the publication listed 42 automobile manufacturers with outputs ranging from 14,000 units a year, in the case of Hindustan Motors, to 9 million units a year in the case of Toyota (*Automotive News* 2007d). This is not to negate the pressure for companies to consolidate into groups, but it does suggest that the accepted theories concerning consolidation lack some fundamental understanding of the forces driving the industry.

1.3 Defining the automobile industry paradigm

Received wisdom has argued that the existence of economies of scale in the industry mean there is an irresistible trend towards consolidation such that smaller manufacturers would either have to join together or else leave the industry. Yet this convergence on a few dominant manufacturing groups is not inevitable, and there have been instances in other industries where firms have exploited economies of scale by limiting themselves to joint ventures that protected the independence of each party. Furthermore, some recently formed large groups have subsequently demerged, DaimlerChrysler being notable here. This indicates that company structure, in terms of the processes and functions that a company owns, is distinct from the question of the scale that these processes should have. If scale is distinct from structure, and given that the production technology dictates the economies of scale, then some advantage may be found in restructuring the vertically integrated structures. This book sets about doing this by assessing the evolution of the industry.

The discussion will aim to clarify the structure of the industry by demonstrating that mass-market automobile production is technically separate from other manufacturing processes due to the unique characteristics of the production technology. I will describe how the consequent economies of scale, which imply synchronizing output between related production processes, such as engine production and final assembly, are well documented in the literature. However, the existence of economies of scale does not in itself compel an organizational structure; but in this book I will consider scale and corporate structure as conceptually separate.

The suggestion here is that just because two separate processes need to match their outputs at some common level for both to exploit the prescribed economies of scale, it does not necessarily follow that both processes should be owned by the same firm. The fundamental question being asked is whether a firm should make a particular part itself or buy it from an external supplier. Since the range of methods for making steel vehicles is the same for all companies in the industry they must all be facing the same make-or-buy decisions, suggesting that they would all come to the same decision on owning the process and so culminate in the same organizational structure. Convergence on a single form of organizational structure is not explicitly articulated in the existing literature and this book analyzes the make-or-buy decision that motivates vertical integration in terms of transaction cost analysis, as espoused by O. E. Williamson. Transaction costs take into account all the costs associated with dealing with external suppliers, whether they can be expressed precisely in monetary terms or not.

1.4 The automobile paradigm and the full-function model

Having introduced the two themes of this research – scale and structure – the purpose of this book is, then, to propose a model for an automobile firm in terms of both its output and its organizational structure. Figure 1.1 shows how I will take these two separate and distinct theoretical approaches to

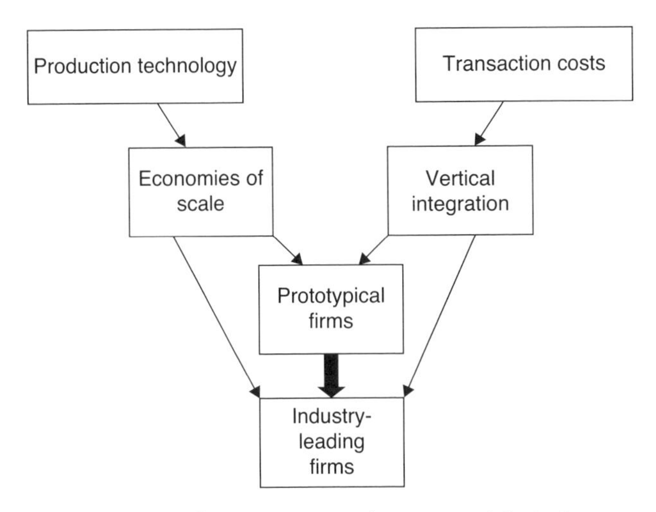

Figure 1.1 Fundamental arguments to the automobile industry paradigm

demonstrate how two types of firm, the prototypical and the industry leader, operate within the paradigm for the automobile industry. The prototypical firm is one that embodies all the qualities of scale and structure set out by the paradigm. An industry leader, though, goes beyond this by offering additional advantages due simply to its dominant size.

From this study it will be possible to derive a systematic understanding of the shared theoretical qualities of automobile companies before they are adapted to fit their own specific contexts. I will suggest a generic conceptual framework for automobile companies by constructing an automobile industry paradigm. The paradigm includes the main economic factors influencing the industry and so provides the defining features of an automobile company. It therefore describes an automobile company that both comprises all the requisite corporate functions and is also of a size to exploit the available economies of scale in each of these functions. I will then take a practical look at how real automobile companies compare with the theoretical precepts of the automobile industry paradigm and the extent to which they lie within the parameters of the paradigm. Of particular interest will be the options for companies that are wholly uncompetitive. The research considers how they might attempt to approximate to the paradigm using alternative organizational structures. The book thereby will encompass three research issues:

1. Optimal automobile company size with reference to economies of scale.

2. Optimal automobile company organizational structure in terms of vertical integration.
3. The degree to which uncompetitive firms can approximate to the optimal size and structure.

In Chapter 2 I will establish the basic principles of economies of scale so that, in Chapter 3, I can then apply them to the automobile industry. This entails studying each of the main functions that are needed for bringing automobiles to the market. For the modern mass-market automobile the defining characteristic is the steel load-bearing body, for which the associated production technology is the body-in-white (BIW) process. The other processes inherent to automobile production are R&D, powertrain manufacturing, and final assembly. It is the central role played by the BIW process that has led to it being termed the Budd Paradigm (Nieuwenhuis and Wells 1997, 2007).

To reveal the prevailing economies of scale in the industry, I will use the technique of Stigler's "survivor analysis." In essence, this approach states that the longest surviving firms in a given industry are the ones that have been able to exploit all the advantages due to economies of scale. Any firm that is in this position is operating at its most efficient and no competitor can beat it on a production cost basis alone. In this book, the research categorizes total global output into output ranges in order to detect any change in the share of the total for each output range. In a stable industry this would demonstrate how firms that produce more than 2 million units a year, for example, are gaining in number and so taking a greater share of the global production total.

Despite the elegant logic of the survivor analysis approach, the results in this research do not conclusively define the economies of scale in the industry, but they do indicate two broad trends. Firstly, that the firms with an output range above 2.5 million units are continuing to expand their share of the global total. Secondly, that although the share of the total held by firms in the 0.5 to 1 million range is shrinking, there are manufacturers within it that have survived for many years and are even growing in absolute terms. It seems that this apparent polarization in the industry is due to the continuing expansion in global output which indicates that the industry has yet to stabilize.

To arrive at a more definitive answer, the research then takes a different approach for uncovering the economies of scale. The minimum efficient scale (MES) for an automobile firm as a whole is estimated by calculating up from the minimum efficient plant size (MEPS) for each contributory process. The data were gathered from archive and company sources for those firms that the survivor analysis had previously identified as showing longevity. For added contextualization, the data gathered by this research is then compared with the data found in the literature or gathered directly from company sources in the UK and Japan.

Taking into account the trend for the very largest companies to expand their market share, the results indicate the presence of two basic types of company. The first is a company with an output of around 600,000 units a year that exploits, as far as is realistic, the available economies of scale. This might be termed the prototypical firm since it forms the basic unit of the industry, and an example of which might be Subaru. The problem for such a company is that its relatively narrow product range leaves it exposed to fluctuations in demand since it cannot offer the disaffected consumer any alternative products. Overall, then, it is unable to achieve consistently maximum use of its plants, and the fluctuation in output leads to higher costs.

The second type of firm is one that has passed beyond the prototypical state and taken a strategic course for expansion to become an industry-leading firm. Although larger firms can suffer diseconomies of scale, due to rising R&D and administrative costs, their higher output and wider model ranges are found to diversify risk and reduce the effects of fluctuations in output volume. Like any firm, an industry leader will still suffer from declining popularity of aging models, or even the abject failure of a new product, but there is a much greater likelihood that it will be able to offer the consumer an alternative and thereby maintain a reasonably consistent production output.

Figure 1.2 shows how the prototypical firm enjoys the lowest costs only if it can keep production of its few models within a narrow output range in its half dozen plants, any variation leading to a steep rise in costs. Alternatively, the industry-leading firm does not, even when operating at its most efficient, have the lowest costs in the industry, but the inevitable

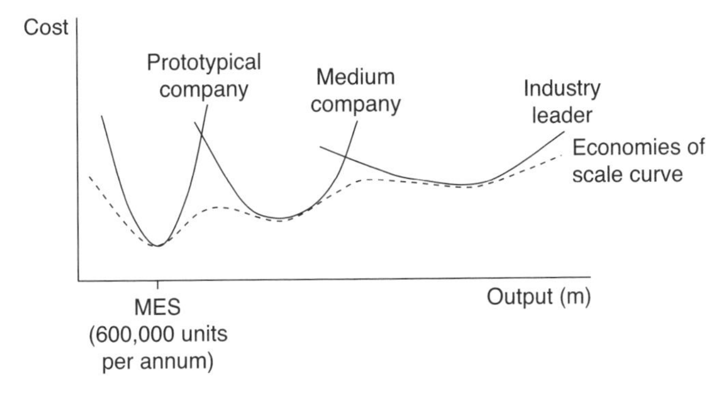

Figure 1.2 Company cost curves and industry scale curve

fluctuation has proportionally less impact. The effect of these fluctuations has been diversified away due to the myriad of products being manufactured in a wide number of flexible plants.

While this provides a guide to the problem of scale it does not specify the extent to which the different manufacturing and engineering functions should be internalized within a single, unified firm. The MEPS found for each plant process does not necessarily compel the vertical integration of the plants within a single company's boundary. This is irrespective of the conclusions concerning the MES for the total production system, which refers to all the production processes, not a single firm per se. On the basis of scale alone there is no reason why the different processes, being technically unrelated to each other, should not be sourced from the supply industry. For this reason, it is not possible at this stage to define which company functions should be included in the vertically integrated automobile industry paradigm. To do this, it is necessary to evaluate the extent of the vertical integration of a paradigmatic automobile company using a different theoretical approach.

Whereas Chapter 3 examines the output scale of the different operations, defining the organizational structure requires an understanding of the principles of vertical integration that might be common throughout the automobile industry. This would dictate which elements, or processes, should be included in the proposed automobile industry paradigm. Corporate organization is discussed in Chapter 4, which investigates the structuring forces within industry. There I will follow O. E. Williamson's use of transaction cost analysis (TCA) to examine the dichotomous issue of whether a company should buy a product on the market or make the product itself in an internal company function. This is known as the "make-or-buy" decision.

Internalization of functions results in vertical integration and in Chapter 5 of this book I will propose a model of comprehensive integration, terming this structure the "full-function model." This is distinct from the term that is often heard, "full vertical integration," which has no predefined limits. The full-function model is illustrated in Figure 1.3, depicting internalization of the core activities related to automobile manufacturing, including R&D, BIW, powertrain production, and final assembly. Information from the market then feeds back to R&D in order that new models can be developed.

The cost advantage of internalization is that it reduces the economic friction of market-based relationships between the functions, which is often experienced in the form of opportunistic behavior by participants as they exploit an unforeseen advantage. For example, a components firm might use its power as a sole supplier to extract a higher price than was

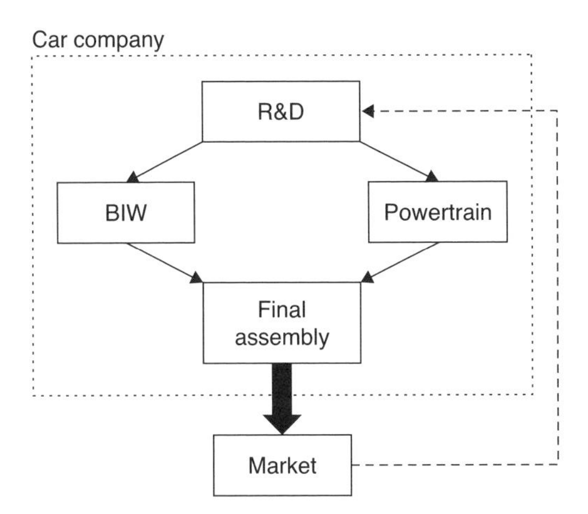

Figure 1.3 Typical full-function automobile producer

originally agreed. The risk of this happening means that it is not possible simply to engage an external firm, it is necessary to incur the cost of putting all manner of contractual checks in place to ensure that the deal goes ahead as intended. Even then, no contract can anticipate all eventualities, so the risk that the other party may exploit a weakness in the contract for their own benefit can never be completely eradicated.

Klein (2000) has demonstrated that this opportunism is primarily due to the human actors, represented at company level by the senior management. If the external company is internalized then any selfish tactics are effectively neutralized since the benefits accrue within the boundary of an integrated firm. However, those individuals operating inside the firm may still have their own self-serving agendas and this has implications for corporate governance structures. The simplest is the U-Form, which centralizes power with a senior manager who has operational and strategic control over the company. Alternatively, Chandler's concept of the multidivisional M-Form of corporation describes a large firm that has a specialist headquarters in charge of overall strategy, which is served by divisions that make their own operational decisions.

In this research the different governance forms are viewed as coping mechanisms for managing the integrated corporate structure. They are therefore management control structures that are overlaid on the full-function organizational structure of the firm, and they are not, in themselves, forms of organizational structure. Indeed, they appear to add layers of management which accrue additional costs and take the larger firm further into diseconomies of scale. This is not to say that larger firms are inevitably less efficient in their administration, but the separation of a specialist strategic control function implies an exploration of economies of scale in senior management which may not be at all clear.

Leaving the issue of governance to one side, the considerations of economies of scale from Chapter 3 can then be inserted in the full-function model of the automobile industry to produce a theoretical size and structure for an automobile company. The discipline of the resulting corporate format is such that I consider it to represent a paradigm in automobile manufacturing, known simply as the automobile industry paradigm. The MEPS (plant size) and MES (overall production system size) data that was illustrated in Figure 1.2 is thereby combined with the full-function structure shown in Figure 1.3. This results in a model for a prototypical firm which then forms the basic unit for the industry, as illustrated by Figure 1.4 below. Since industry-leading firms expand by the addition of complete plants the firms can be considered to grow by units of MEPS and, in combination with, MES. They can therefore be perceived as multiples of the prototypical firm, albeit with the danger of higher administrative costs which then take the firm into diseconomies of scale.

However, for smaller firms that have not yet reached the prototypical state, the first priority is to construct plants that can achieve this overall optimal output of MES, assuming the market demand is sufficient. It is not possible for a small firm to grow in units of MEPS since the firms have

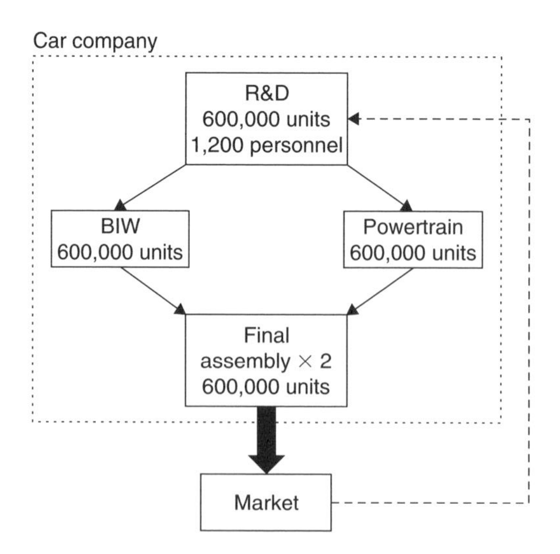

Figure 1.4 Automobile industry paradigm

yet to achieve even this level of output for the plants they do operate. Until they have expanded their output to the minimum level of production they will suffer a cost disadvantage in comparison to the prototypical firms that have achieved MEPS for each function and MES overall. Given that this will take a period of time that may be beyond the capabilities of the firm, an alternative in the short term would be to implement a strategy that approximated to the size and structure advantages enjoyed by prototypical firms within the paradigm. Chapter 6 of the book therefore discusses the range of options for approximating to the automobile paradigm. These include exploiting economies of scale through life extension of the product cycle and the more intensive use of the production facilities through flexible manufacturing – although these are both strategies that are available to larger manufacturers. Alternatively, from a structural point of view, an uncompetitive company could achieve lower structural costs if it were able to find a novel form of organization that could replicate the cost advantages of vertical integration. This means forming a relationship with an external company, but not through the external mechanism of the market, since this would carry with it precisely those economic frictions that the full-function integrated structure of the paradigm was able to avoid through internalization. Chapter 6 investigates ways in which the full-function organization can be divisible between partner firms in a manner that accesses the advantages of vertical integration and avoids full internalization.

Dussauge *et al.* (2004) state that two forms of alliance are found to be particularly promising:

- The scale alliance, where partners share functions in order to raise output and exploit economies of scale.
- The link alliance, where partners exchange knowledge, each making a unique contribution.

Scale alliances are found to be long lasting due to the persistent requirement for economies of scale. Conversely, link alliances have been found to last only until the need for knowledge is satisfied or one partner gains an advantage over the other. However, this research uses TCA to suggest that link alliances are more enduring if the partners have ownership of complete and complementary functions. Any learning benefits would then accrue solely to that partner and so eliminate the risk of opportunism where one partner could take advantage of the other. The structure by which this can be achieved is the vertical joint venture (VJV), each partner being assigned to discrete functions which make up the full-function structure. For example, one partner would take responsibility for product development in its entirety, while the other was responsible for all the remaining functions. This is illustrated by Figure 1.5.

Chapter 6 also looks at international differences between potential partners, according to their locations, to show how this can bring additional benefits to a VJV. This focuses on the foreign direct investment (FDI) literature, starting with the work of Hymer and elaborated on by Kindleberger. This is concerned with factor endowment differentiation, which is the manner in which foreign locations can offer distinctly different resources. For example, some developing nations have large pools of low-cost labor. Schott (2003) argues that these differences evolve over time, particularly regarding the growth in wage costs. For a more dynamic aspect, Vernon's product cycle and stage theories of progressive internationalization are introduced. Taking the various theories to be mutually exclusive, Dunning draws their aspects together within the wide ranging eclectic theory, which covers international opportunities in ownership, location, and internalization (OLI).

Strategic theories of FDI, as put forward principally by Cowling and Sugden, suggest that firms may engage in FDI in order to enhance their competitive position relative to rival firms. I have indeed found that firms will expand output capacity in order to diversify risk, much of this expansion taking place overseas. However, FDI is only incidental to this strategy, since the same result could be achieved within the home market if it were

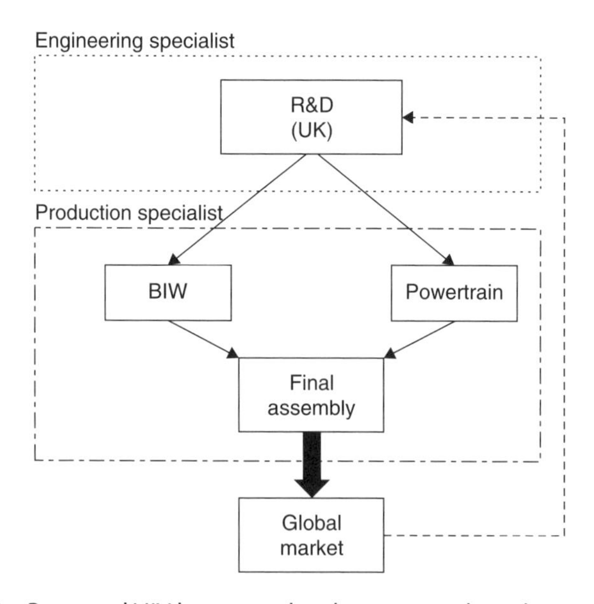

Figure 1.5 Proposed VJV between development and production partners

large enough to support the output volume. Although strategic theories of FDI are informative, my research is interested primarily in the underlying economic factors that impinge on the industry and, as such, it is these that remain the focus of the study. Instead, factor endowment differentiation can be applied to the VJV structure, such that the respective partners take responsibility for those functions which are already endowed with the relevant factor advantage. To take a specific example: this might result in an American company retaining product development in the US while moving assembly work to India. This discussion concerning the added advantage of selective relocation culminates in the proposal for a structural and global division of the full-function model in the form of an IVJV.

The IVJV is particularly pertinent to the demise and partial resurrection of the British mass-market automobile manufacturer, MG Rover. Chapter 7 charts the decline of MG Rover's predecessors, BL and Rover Group, to provide a historical perspective on MG Rover's condition during the period 2000–05. The reasons for the ultimate demise of MG Rover in 2005 are discussed along with the partial failure of the company to enact an IVJV. As an adjunct to the collapse of MG Rover, in Chapter 7 I also discuss the manner in which NAC and SAIC, two Chinese automotive manufacturers, purchased the respective physical and human assets of MG Rover and so pursued variations of the IVJV structure, thus replicating the advantages of the automobile industry paradigm. More recently the two companies have merged, still without undermining the basic format of the IVJV. For clarity, the proposed IVJV, that might have evolved if the alliance between MG Rover and SAIC had gone ahead, is presented in Figure 1.6.

The final chapter in this book, Chapter 8, draws together all the research findings to show how the automobile paradigm challenges the established perspective of the industry. Although output size seems to have become the mantra for the industry, the research indicates that this is inappropriate and that many companies have found that consolidation simply means that they are effectively multiple automobile companies, each needing to pursue the economies of scale by itself. Such multiple full-function firms simply multiply the problems they face.

Ford built just such a multiple full-function structure as it subsumed a clutch of foreign brands. The company has since withdrawn from its strategy of product diversification with the divestment of Aston Martin, Jaguar, and Land Rover so that it can now focus its resources on its traditional heartland, the mass market. Similarly, other large firms have found that their inflated profits from a successful core range can mask inefficiencies elsewhere in the organization. Should the competitive landscape change then these firms might come to regret dissipating their resources

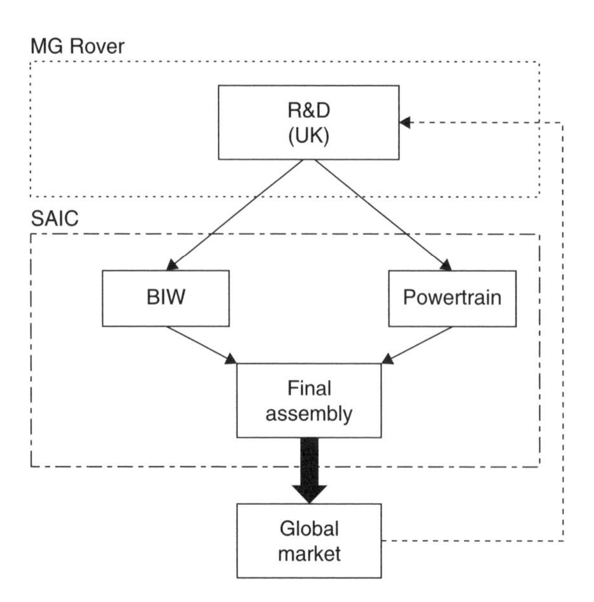

Figure 1.6 Proposed IVJV between MG Rover and SAIC

in such adventures as robotics, aeronautics, and fanciful concept cars. By contrast, Toyota may now be the world's largest automotive producer but it seems to be able to retain for longer the cost advantages of smaller firms by breaking itself down into a network of quasi-independent units. Indeed, the majority of products manufactured in Japan under the Toyota brand are in fact made by affiliated companies in which Toyota only has a minority holding. This seems to create an element of internal competition that can preclude opportunism and loss of corporate focus.

For those firms that exist below the prototypical benchmark of a strictly limited product range manufactured at a rate of 600,000 units a year, the inherent cost disadvantage can only be offset in the long term by external aid, most probably in the form of government support. However, in a free trade regime such aid is prohibited on the grounds that it bestows an unfair advantage on the firm. Normally, full exposure to the global competition would compel a mad rush to reach a world class standard or risk being driven from the market place. Given the brief time available, the IVJV represents an instant alternative for achieving sustainability. The IVJV concept particularly lends itself to firms in emerging economies that can call on substantial production resources but lack abilities in technical development. As I point out in this book, China's SAIC is pursuing just such a strategy in reconstructing the assets of MG Rover. In such cases, the paradigm suggests that the firms should actively promote the international division of their organizational structure in order to exploit the contrasting advantages, even if it means relinquishing control of functions they once held dear.

Once firms have reached the prototypical state, whether alone or as part of a VJV, they enjoy all the cost advantages of being the most suitable size and structure for the current state of the industry paradigm. It must be cautioned, though, that they are vulnerable to rapidly rising costs whenever output is allowed to fluctuate from the optimum. The defense against this is to dampen these fluctuations by devising a range of products that is highly differentiated but conservative in the long term, resulting in stable demand. Thus, a firm such as Subaru should remain committed to its unusual engine technology, four-wheel-drive systems, and distinctive body style. If by such a strategy similar manufacturers can survive, then this should result in greater diversity in the market as each is obliged to find a unique niche to defend.

As for the automobile paradigm itself, like all paradigms it will continue to be refined until it is superseded, but not directly replaced, by a competing paradigm. This is surely likely to happen in the coming decades as oil-based fuel is phased out. If it is true that fuel-cell power represents the next great leap forward in the industry then it is crucial to recognize that this is not a technology that will simply supplant the traditional internal combustion engine. Since the packaging of fuel cells is entirely different to that of the traditional engine it will be possible to design and manufacture vehicles in a distinct new form. This is most clearly demonstrated by GM's "skateboard" platform concept that separates the chassis/frame from the vehicle body. Like a return to the pre-Budd era coachbuilders, such a system might result in vehicle manufacturers being high volume suppliers of platforms to small, localized firms that fit bodies of their own design. This will signify a new paradigm in the organizational structures of firms and the economies of scale that dictate their size.

Yet despite the publicity surrounding fuel cells and other new technology, the future is far from certain. If the new developments are to have any application to personal transport, then the costs and reliability of the technology will have to be sharply lowered. The risks involved are so high that one might even wonder if commercial firms have any business straying outside of their paradigm when there is still so much work to do within it. Indeed, it could be argued that blue skies research into new paradigms should be the preserve of national governments. In the meantime, there is plenty of scope left in the current automobile paradigm for further technical refinement and thus a promising future for automobile manufacturers the world over. The automobile industry as we know it is shaping up to dominate the twenty-first century just like it did the twentieth century. If globalization is a rising tide, then there is no reason why it should not raise the entire automobile industry with it.

CHAPTER 2

Fundamentals of scale

In the course of defining the generic structure of a mass-market automobile manufacturer it is necessary to establish the forces that form the company. These give the firm the structure and size of its operations. The organizational structure of the firm is defined by the degree of vertical integration, involving the decision whether to source production externally or produce internally, known as the make-or-buy decision. The size of operations is defined by the available economies of scale which encourage the firm to adjust output to a point where costs per unit are minimized.

Different industries will have widely variable integration and scale considerations. There can also be variety within industries as different firms utilize resources in their own unique ways. However, if an industry has a defining technology then all the firms operating that system will be drawn into the same set of economic conditions. The firms will then converge on a generic firm structure.

This chapter of the book will illustrate the underlying economic considerations of scale and then demonstrate how they impinge on the automobile industry. As a multiproduct production process, including major components such as engines, transmissions, and steel load-bearing bodies, automobile manufacturing comprises a complex array of industrial activities. Each activity has its own optimum level and these need to be matched in order that the total production process should exploit the available economies of scale. Beyond the production processes there are also economies of scope which bring additional advantages to a firm when adding new products to its range, thereby utilizing existing company facilities more efficiently. Although economies of scope are not derived from the production technology, and therefore do not determine the economies of scale in the industry or the ideal size of firms, by raising overall efficiency they can engender greater sustainability.

2.1 Economies of size

Large firms enjoy a number of benefits over smaller rivals: Pratten (1971) pointed to the power that comes with market dominance. A firm might find it has some monopoly power in setting market prices or extracting lower prices from its suppliers. A larger firm might have political power, able to influence legislation and tax revenue policies in its favor. Internally, larger firms also benefit from team production where members of the team can specialize in tasks (Parkin *et al.* 1997).

The two main advantages that accrue to a large company are connected with its production capability: economies of scale and economies of scope. Economies of scale are related to technical efficiencies, and industries that experience its effects find that unit costs of production decrease as plant size is increased. It is important for planning purposes to note, however, that beyond a certain point an increased size of plant may be less efficient; progressively larger plants are then said to be experiencing diseconomies of scale. This consideration of optimal size is a core theme of this book since economies of scale indicate the ideal plant size for a firm, and thereby the ideal size for the firm as a whole.

Economies of scope exist when a large firm gains efficiency by more intensive use of its facilities. Although economies of scope are not prescriptive in structuring the firm, and therefore not core to this book, they do indicate where a firm might improve its sustainability. For example, the cost of selling an additional product through a shared distribution system might be less than for a rival selling the same product through a dedicated distribution system. The product itself can also benefit by sharing components or even the specialist knowledge in R&D.

Overview of economies of scale

Cairncross (1966, 106) described two different areas where scale may be found: external economies and internal economies. External economies occur when firms have grown large enough that the industry can support its own infrastructure, made up of, for example, specialist support and trained labor (Sloman 2001, 95). Cairncross (1966, 107) also included economies of concentration – where firms congregate in a single location, a tendency recently promoted by the growth of just-in-time (JIT) methods (Womack *et al.* 1990). Economies of concentration, then, inhibit geographic spread and promote vertical integration by introducing cost advantages when companies are in close proximity. The decision of a firm to integrate vertically or not will be examined in later chapters. The converse – economies of disintegration – concerns the breaking down of the production process into specialist functions that are better served by separate firms or even industries. This does not necessarily promote global spread but it does enable it and suggests opportunities for divisibility within the production transformation process as a whole. Divisibility of production will be covered in Chapter 6, with specific reference to vertical joint ventures.

Internal economies of scale, on the other hand, are specific to the firm and Cairncross (1966, 108) listed five of them:

- Financial economies: the ability to raise capital more cheaply with increasing returns to scale, so that as the firm expands these advantages show a proportionally larger increase.
- Marketing economies: increasing returns to scale through a mix of growing monopoly powers and increasing use of sales capacity.
- Risk bearing economies: reducing exposure to risk by expanding with increasing returns to scale, though a firm can structure itself to similar effect by diversifying output, markets, and supply.
- Managerial economies: manifested through delegation of detail or functional specialization. Delegation only comes with size, management having its own indivisibility, though beyond a certain size there may result an unwieldy bureaucracy and therefore diseconomies of scale.
- Technical economies of scale: based on production technology.

Technical economies of scale particularly relate to plant size, though the manner in which the technology is used can still distinguish one firm from another. Cairncross (1966, 109) put forward three ways in which a firm may utilize the technology to its own advantage, resulting in economies of superior technique, increased dimensions, and linked processes. Dimensional advantages are clear in, say, a blast furnace where a doubling of its dimensions will increase its potential internal capacity eight times. In contrast, there are instances where there are clear limits to dimensional advantages if they are held back by the nature of the product, such as furniture manufacturing, where there are both constraints on the natural size of wood supplied and the practical dimensions of the final product.

There is also a lower limit to what might be considered to be a production system. Cairncross (1966, 106) stated that there is a qualitative difference between the lone worker and the mechanized production system. To work on a variety of tasks the lone worker must be a generalist, able to vary the intensity and allocation of the work but not be able to delegate functions. At larger volumes a production system must be put in place by an organization, with specialization in tasks by workers and machines. This is particularly characterized by indivisibilities: the inability to break a task down into smaller units.

A plant of a certain size comprising a specified production system can produce for a whole range of outputs but it will be at its best when all the factor inputs are working at their most efficient. This pace of work will be when the average unit production costs are at a minimum, this being the point where the production system is operating as quickly as possible but without affecting the quality of the product or requiring disproportionately higher factor inputs. For example, a bread-making factory is at its most efficient when the conveyer belts are fully loaded and working at the required speed for the bread to be properly baked. In the short run, demand for loaves is likely to fluctuate and it will be necessary to adjust the production volumes. However, to attempt to increase output by raising the temperature of the ovens or accelerating the speed of the conveyers would reduce cooking times and ultimately lead to more loaves of sub-standard quality being rejected. For those loaves that did make it through, the average cost of baking them would actually rise. If the market demand for loaves slackens, then the number of loaves being baked can be reduced, but since the ovens must be kept at the same temperature the average cost of production for these loaves is also higher. In such a scenario, the only input cost that can be varied is that of the dough itself, all other costs being fixed.

For production systems in general, the most efficient production levels may represent a comparatively narrow band of output compared to the full range that the system can physically operate at, or the minimum average cost may even be denoted by a single level of output. It is at this point that the specific plant is being operated at the most efficient level and this denotes the optimum output level for the plant, i.e. the level of output that generates the minimum short-run average cost. Either side of this point average costs are higher, either because the plant is underemployed or because it is being overworked and diminishing returns to the various factors set in. The firm may accept this if it achieves the planned output, but it should be aware that average costs are not at their minimum. Our bread-making firm would probably accept that production would have to vary with demand as long as it does not stray too far from the point of greatest efficiency. An alternative tactic might be to hold production constant at the most efficient level and then vary the price in order to maintain sales, discounting being the most common approach. It would be up to the firm to decide which of these short-term tactics were most appropriate, the only option not open to it is to change the fundamental production characteristics of the plant since these are fixed.

This is not the case in the long term. Here, the company can plan for a plant that provides the level of output considered appropriate for the demand it believes exists for the product. For this hypothetical plant the company is free to explore all variable and fixed costs and arrange them as it sees fit. For example, it might decide that production will remain fairly constant and so will decide to invest in machinery. This will have a high fixed cost once it is installed, but this is not a problem as long as the machinery is run at its most efficient rate. Alternatively, the firm might prefer to employ more labor in the belief that employee numbers can be readily adjusted to suit the variability in demand. In either case, it is assumed that a larger plant will offer lower average costs, otherwise the smaller plants could simply be multiplied. It is also assumed that as the scale of the plant increases, drawing in greater factor resources, the relative factor prices will not be affected (Pratten 1971). In practice a very large firm will indeed distort factor prices, particularly by raising local wage rates, but for clarity this is assumed not to happen at the theoretical level.

Of course, once a particular plant has been decided upon then only certain inputs can be readily varied, even as part of a planning exercise. Each plant has its own short-run average cost curve, but, for planning purposes, when all possible plant cost curves are plotted together, then the long-run average cost curve traces a line around all of them, often being referred to as the envelope curve. Taking a long-term view, a company can select the most appropriate plant by inspecting this curve and deducing which size of plant is adjacent to the targeted level of output. Indeed, terming it the long-run average cost curve is misleading since it is not intended to show how costs change over a period of time and Pratten (1971) preferred to label it the scale curve. Nevertheless, the long-run average cost (LAC) curve is the generally accepted terminology. The optimum plant size is the one whose minimum short-run average cost point coincides with the minimum for the long-run average cost curve and is thereby able to exploit all the available economies of scale. This is known as the MEPS.

Although economies of scale are available for almost any process, the search for the technical optimum associated with production – the output level either side of which would incur cost disadvantages – has received most attention because it renders itself amenable to quantifiable analysis. If the technical economies of scale are very large they become a dominant feature of the firm and give the impression that this is the source of the firm's competitive advantage. However, this may mask a lack of economies of scale elsewhere, even diseconomies, particularly in a multiplant firm where a complex bureaucracy may result in diseconomies of scale in management.

A role for economies of scope and integration

Economies of scope are often cited as an additional advantage, usually exploited by large firms that have a wide range of capabilities and resources to draw on. These economies of scope are available when the production system is established and a more intensive approach is made to its usage. In order to exploit the opportunity the minimized costs of production for two goods would have to be less when the system is shared than if the goods were produced separately (Panzar and Willig 1981). Typical examples include the production of mutton and wool, or a delivery person who also reports on neighborhood crime.

The essence of economies of scope is that it accesses additional capacity derived from an existing production system with the result that total output is improved. However, the actual advantage can be difficult to quantify and may in fact represent increased use of existing spare capacity, rather than the opportunity to exploit additional capacity. Since it is derived from an existing production system it is also likely to have only a marginal impact on raising total production levels. For example, there are no economies of scope in a food delivery firm diversifying into mail delivery since it is only possible to do this if the company is failing to entirely fill its vehicles with food deliveries in the first place.

For the increase in scope to bring economies there must be spare capacity available which cannot be used for the existing product. As Clarke (1987, 113) described it, the inputs should be used without "complete congestion." Such inputs are quasi-public in that they have characteristics in common, as opposed to private inputs which are mutually exclusive. If the production of one good squeezes out the production of another, to the detriment of the cost structure, then economies of scope are not being exploited. Equally, shared production does not mean that economies of scope exist since it might simply be a case of filling otherwise redundant production capacity. Alternatively, exploiting economies of scope may not simply add to existing scale but lead to its increase, as when an R&D function is expanded to allow the engineering teams to take on a wider range of projects. This might result in diseconomies of scale in R&D which are compensated for by economic advantages elsewhere, such as economies of scale in production or economies of scope in the product range.

Where economies of scope exist they can have integration implications for the firm (Panzar and Willig 1981). Multiproduct firms that are supplying goods which have quasi-public inputs have a motivation to integrate in order to access the economies of scope. Credit unions, for example, enjoy economies of scope in supplying a range of financial products and so tend to integrate them (Youn Kim 1986). However, it is not clear how much of the resultant savings are actually due to economies of scope. If it means that the same personnel can sell a range of financial products, pensions, and insurance, for example, then the advantage comes from making greater use of the sales personnel. The alternative would be to employ specialist staff who would find themselves idle while they waited for their customers, unless, of course, demand was high enough to keep all the specialists busy. This, then, is about capacity utilization rather than economies of scope.

Applied to a multiproduct firm with separate functions the emphasis on either scope or scale is delineated only with difficulty. To take, as an example, furniture production: the two manufacturing processes for the wooden frame and the soft coverings are mutually exclusive but the designs are sourced from a common R&D function. The design function has the capability to work on many different styles of furniture, from sofas to sideboards, the knowledge in one being relevant to another. This suggests that economies of scope can be exploited, but for designers to work on one project they must abandon another. Economies of scope would only emerge if the designers could develop a technique in, say, sideboard design and apply it advantageously to sofa design – but in a way that could not have been learnt by specializing solely in sofa design. This is unlikely, and the best that can be said is that designers that divide their time between both products are just going to be busier than those that specialize. This is capacity utilization. Where economies of scope occur it might be in reducing the learning time, for example, of a new design technique being more readily learned in sideboard design than sofa design, although the benefits would be marginal.

This view can be extended to the firm's managers, with managers of multiple divisions simply being busier than managers of single divisions. Clarke (1987) cautioned against viewing a firm as if it were one structured by economies of scope for two reasons. The first is that shareable inputs will compel vertical coordination but not vertical integration. The second is that diversifying in order to fill existing capacity comes within the definition of economies of scale. For this reason, if economies of scope in management exist at all, they will be too weak to be decisive in pushing for an integrated company structure.

Returning to the example of the furniture industry, at the core of the firm are the manufacturing functions for carpentry and textiles. These functions will be brought together within a single firm if it is found to be more cost effective to integrate the functions rather than transact with each other as separate entities regulated by contracts. Since the production processes have no relation to each other there is little opportunity for economies of scope. If R&D is included as another function in the production process then all furniture firms, given the same economic environment, will tend towards the same fundamental organizational structure of functions. Whether this results in the integration of these functions or their trading with each other as separate companies will depend on the relative costs of the two arrangements. For the structure of a multiproduct firm, with technically unrelated functions, the integration decision is concerned with transaction costs (this will be discussed further in Chapter 4). I will continue by looking at the size of each function in terms of the output necessary to exploit the available economies of scale.

2.2 Production costs

The economic fundamentals upon which all companies are founded are based on the role of production costs. At the simplest, the cost of production is a multivariable function taking in all the sources of production costs. Koutsoyiannis (1979) expressed this as follows:

$$C = \text{fn}(X, T, P_f)$$

where C is total cost, X is output, T is technology, and P_f is the price of factors of production.

This can be further simplified if it is accepted that costs (C) are a function only of output (X) as long as other cost factors are constant. If this were represented graphically, then changes in costs due to output would be found by reading along a cost curve relating X and C, whereas changes in the hitherto constant factors would be shown by a complete shift of the cost curve. Technology (T), for example, includes notions of entrepreneurial and technical innovation resulting in a change in the type or method of production, therefore resulting in a shift in the cost curve.

When considering the long run, all costs may be varied, and so they inform the management as to how to plan for the future. The costs are calculated *ex ante* and are used to consider the optimal production system for an intended level of output. Long-run cost curves include the firm's internal economies of scale, while external economies of scale are beyond its control and so changes there, inducing changes in factor prices, will bring about shifts in the long-run cost curve. In the short run, however, the production system has been selected. The management is now faced with the fixed costs of the production system that it has put in place, comprising the capital equipment which cannot be changed in the short run and the variable costs over which management does have operational, short-run control.

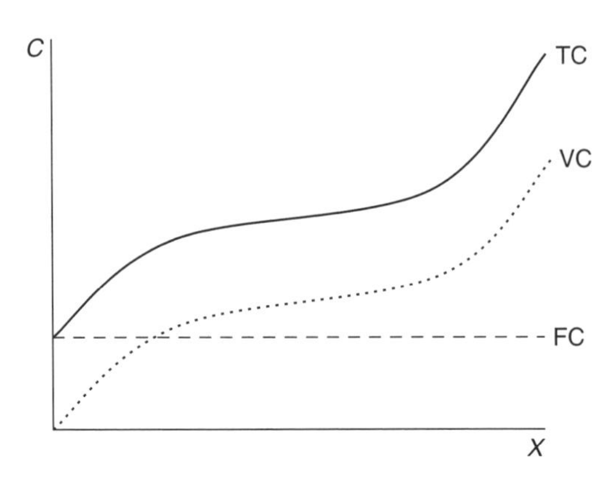

Figure 2.1 Fixed cost, variable cost, and total cost in the short run

Short-run costs

Total costs (*TC*) of production in the short run are assumed to be either fixed costs (*FC*) or variable costs (*VC*) (Sloman 2001, 86). *VC* are costs that vary with output volume in the short run, taking in supply materials, direct labor, and fixed capital running costs. *FC* are taken to be those costs that are not amenable to being varied in the short run, such as depreciation, fixed maintenance schedules, and administrative salaries, even if they are variable in the long term. *TC* is therefore described by the formula:

$$FC + VC = TC$$

Initially, variable factors of production exhibit an increase in productivity as more are employed, as a result of increases in marginal products, resulting from increasing returns to the variable factor. Beyond a certain level of employment, however, inefficiencies develop and diminishing returns to the variable factor set in. *VC* can therefore be represented by an S-shaped curve (see Figure 2.1), while *FC* can be represented by a straight line parallel to the output axis, the two representations being summed to arrive at the *TC* curve.

When looking at the average cost per unit of output, the S-shaped curve for *VC* translates into a U-shaped, short-run, average, variable cost (SAVC) curve (see Figure 2.2). The initial increase in productivity is expressed as a fall in cost per output or average variable cost until a minimum is reached at the trough in the U-shape at output X_1. From there, diminishing returns are experienced and so the SAVC curve shows a positive gradient. Since *TC* is the sum of *FC* and *VC*, the short-run, average, total cost (SATC)

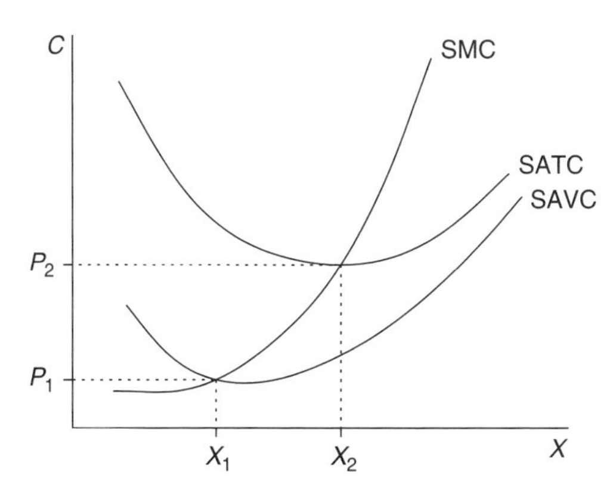

Figure 2.2 Plant cost in the short run

is derived in the same way as the SAVC and so similarly results in a U-shaped curve. The SATC, therefore, shows the costs of running the plant as a whole for different levels of output. In studying these curves, management are able to identify the point of minimum costs per unit of production (X_2 in Figure 2.2).

The addition of average fixed costs (AFC) to the SAVC has the effect of shifting the SATC curve up and to the right of SAVC. Close to the origin, *FC* has a large impact on *TC*, so the SAVC and the SATC are some distance apart. As output increases due to the greater use of variable inputs, AFC falls and the two curves converge. The relationship of the two curves to the short-run marginal cost (SMC) is such that the SMC intersects the SAVC and SATC at their lowest points, X_1 and X_2, respectively. The point X_2 shows where the firm is operating at its optimum for the whole plant, i.e. the minimum of the SATC. Parkin *et al.* (1997, 265) described how the intersection of the SMC and SATC represents a point of equilibrium where the firm is capable of breaking even, or earning "normal profit" as the authors refer to it. In terms of revenue, below P_1 variable costs are not being covered, so there is no rational reason for production. For any unit price above P_1 and output X_1 the variable costs are covered and a contribution made to the fixed costs. At price P_2 and output X_2 all the costs, both fixed and variable, are covered and the plant earns "normal profit." On the SATC curve any output above or below the optimum denotes higher average costs. In the long run this may put the company at a cost disadvantage in comparison to competitors, so it will need to plan a resizing of the plant. This is covered in the next section dealing with long-run costs.

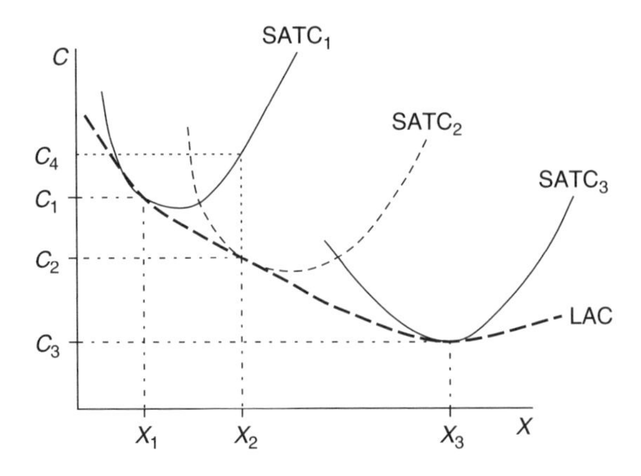

Figure 2.3 Alternative plant sizes for long-run planning

Long-run costs

As has been noted previously, in the long-run view all factors of production can be varied because the firm has yet to construct a specific plant. This means complete freedom in the planning of facilities for future production as the firm can look at different plant sizes. According to Pratten (1971, 4) the question facing the firm is as follows: "What would the effect of scale be on the average costs of production of a series of alternative plants built at a point in time, each perfectly adapted to the required scale and operated at that scale?"

The firm is able to consider different plant sizes, each with its own distinctive production systems, based on the current technical knowledge. Rosegger (1986, 72) cautioned that the LAC curve does not describe costs in the long run but reflects the cost structure of a long-term commitment and is therefore a planning aid. The plants are theoretical at this stage but can be realistically costed, *ex ante*, as part of the planning process. Figure 2.3 shows the SATC for three plant sizes: $SATC_1$, $SATC_2$, and $SATC_3$.

The LAC envelopes the SATC curves for the three plants under consideration. It is drawn as a smooth line on the assumption that there are an infinite variety of plants, not just the three shown. If this were not so and there were only three available plant sizes then the LAC would be scalloped as it traced around the outside of the three plant curves, from one curve intersection to the next (Thompson and Formby 1993, 224). The company would then select the plant size that was best suited to the appropriate level of output in the long run.

If output X_1 is planned for then it is clear that the firm would select system $SATC_1$ with average cost C_1. At output X_2 the firm would need to decide if the new output was for the short or long run. It therefore has a choice between staying with $SATC_1$ and accepting that average costs will rise in the short run to C_4, or taking the long-term view and reconstructing around plant $SATC_2$ and benefiting from the lower average costs, C_2, in the long run. The crucial decision depends on the future long-run expectations of the firm. Taking the long view, the company uses the LAC to select the plant which, in the long run, will produce the company's planned level of output at the minimum unit cost, but which can, in the short run, cope with temporary increases or decreases in output.

At output X_3 the company would be aware that plant $SATC_3$ would not only be the most suitable for the desired output but would also give access to all the economies of scale available. Up to this point increases in plant size would enjoy increasing returns to scale, but beyond this there would be decreasing returns resulting in diseconomies of scale. The minimum point for $SATC_3$ coincides with the minimum of the LAC to denote optimum scale in both the short run and the long run. The LAC is also in tangential contact with $SATC_1$ and $SATC_2$, but not at the minimum points for each curve. Assuming that the production systems are inflexible then, for $SATC_1$ and $SATC_2$, the contact is on the negative slope of the SATC curves, revealing increasing returns to scale. In Figure 2.3 this is shown by output X_2 which is achievable, either by underutilizing plant size $SATC_2$ or by overutilizing plant size $SATC_1$, though, as already noted, plant size $SATC_2$ gives rise to a lower unit cost (Thompson and Formby 1993, 225). Those plants that are larger than $SATC_3$ are assumed to suffer diseconomies of scale, perhaps due to the managerial diseconomies described by Cairncross (1966).

The development of long-run costs can also be expressed incrementally by investigating marginal costs (MC). MC in the long run (LMC) is not an envelope curve, since it connects the points on the SMC curves that signify the output where the LAC–SATC curves are in tangential contact. This is clarified by Figure 2.4 which shows the SMC and LMC intersections at outputs X_a and X_m, omitting $SATC_2$ for clarity.

Moreover, X_m marks the intersection of four curves in total such that:

$$SATC_3 = SMC_3 = LAC = LMC$$

The firm will choose whichever short-run cost curve gives it the lowest per unit cost in relation to the desired output. X_m, though, brings further advantages because it is the point at which the lowest absolute per unit cost is achieved in the long run, assuming that the firm can justify the level of

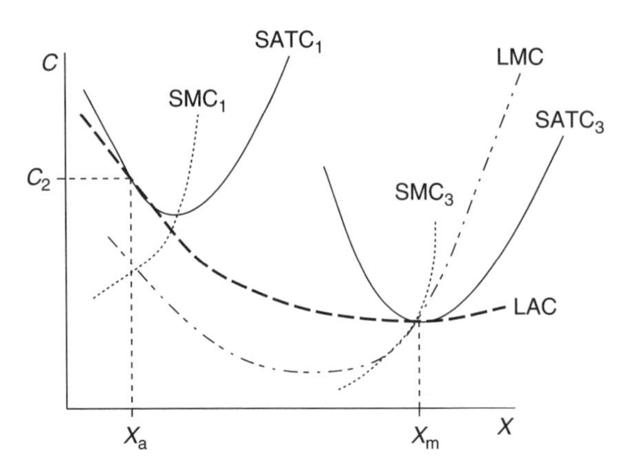

Figure 2.4 Marginal cost in the long and short run

production. This, then, is the optimum plant size and is the one that can exploit all available economies of scale. The smallest size of plant, or lowest output, that still exploits all available economies of scale is said to have the MES or be the MEPS. The terms MES and MEPS tend to be interchangeable in the literature, although I will use MEPS when referring to specific plant sizes and MES for the firm as a whole, implying a number of plants within the firm, each of which exhibits MEPS. In Figure 2.4, the MES and MEPS are represented by output X_m for a single plant firm.

In practice, it may prove difficult to define the MEPS since the planning horizon given by the LAC is based on existing technology, and a cost-minimizing, profit-maximizing firm would not seek to offer itself as an experiment in diseconomies of scale. For this reason plant sizes beyond the MEPS tend to be speculative or exploratory in character, rendering the shape of the LAC highly uncertain after the MEPS. Pratten (1971) did not consider the MEPS to have progressed if a newly developed plant doubles the scale but delivers a reduction in total average unit costs of less than 5 percent and a reduction in value added per unit of less than 10 percent. The next section will look at how different shapes of the LAC affect the long-term plans of companies.

Alternative long-run, average, cost curves

The implication of the U-shaped LAC curve is that there is one size of plant that enjoys all the economies of scale and thereafter increasing plant sizes are weighed down by decreasing returns to scale. In other words, the larger plants suffer from increasing unit costs. However, the danger of this occurring may cause companies to avoid exploring the LAC curve beyond the established MEPS and so the precise shape of the curve remains largely conjectural (Pratten 1971).

Indeed, Thompson and Formby (1993) suggested three basic shapes of the LAC curve. In addition to the classic U-shape they suggested curves skewed to the left or to the right, and another in a flat-bottomed saucer shape. The flat curve means that firms with plants of different sizes can operate with the same level of average cost. This should not be taken to mean that a large plant can be run at the same output quantity as a smaller plant for the same average cost, but it does mean that for planning purposes there is a wider choice of outputs, with each denoting an optimum of a particular plant's SATC curve that is equal to that of another plant of a different size. The smallest size of plant that is still on the flat, lowest, section of the curve is the one that has achieved the MEPS and suffers no production cost disadvantage against any other size of plant on the flat section.

The range of LAC curves is not even restricted to these three types, and neither do they need to show smooth, regular shapes. Since the LAC curve is traced around the SATC curves of the various plants, its shape is dictated by their shape and how they are positioned relative to each other. It is quite conceivable that the LAC curve could take on an irregular scalloped profile if the progressions between plant sizes are not sufficiently uniform.

Integrating different production processes

At the plant level there may be supplementary processes which are indivisible. When integrating different processes the optimum output then becomes the optimum for all the processes in combination with one another. Maxcy and Silberston (1959, 76) gave the simple illustration of two machines, A and B, Machine A producing 25,000 items a year to feed Machine B, which has a capacity for 50,000 items a year. To achieve the 50,000 items, though, would mean duplicating Machine A. Problems occur if demand slackens and an output of only 25,000 is achievable, at which point neither Machine B nor the two Machines A are running at their optimum. At this level, the costs for two Machines A are higher than if the company had planned for the lower level and installed only the one Machine A.

Complete processes are often a compound of manufacturing activities like these. For example, forging involves not only molding the parts but also machining them, followed by assembly. There being no common technology between the processes, synchronizing them can be complex. The principle can be extended to entire processes contained within one plant. The optimum for a company is dependent upon the lowest common multiple of the processes that make up the firm's activities (Rhys 1972, 289).

The consolidated figure for all the different outputs would be the optimum for a single production transformation process contained within a plant. Firms do not necessarily structure themselves around this one process, though, and indeed may encompass technically separate plants which provide the main components for assembly into a final, complex product. At the planning stage, the multiplant company would need to be aware that the short-run cost curves for all the plants would need to be synchronized according to the same principles as for individual machines within one process. Moreover, the company would need also to refer to the relevant long-run cost curves for each plant to arrive at a long-run structure for the company as a whole.

Cost curves for the multiplant firm

The multiplant firm is one that manufactures items using technically distinct processes. If the product is kept constant while the plant sizes are varied, which might happen if the plants are of different ages, then, as Thompson and Formby (1993, 396) recommended, three steps are needed to bring the plants into the most technically efficient alignment:

- Step 1: calculate the firm's *MC* from the various plant *MC*s.
- Step 2: calculate the most profitable output where overall *MC* equals overall marginal revenue (*MR*).
- Step 3: allocate profit maximizing output among the plants so the *MC*s of the last products at each plant are equal.

This can be expressed by the formula:

$$MR = MC = MC_{P1} = MC_{P2} = \ldots MC_{Pn}$$

where $P1$ and $P2$ are various plant sizes up to plant Pn.

In the case of a multiplant firm manufacturing component parts for a single final product then the plants differ not simply in size but in the

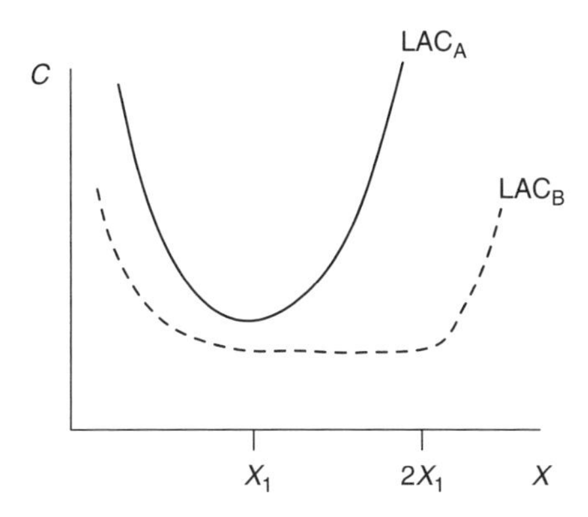

Figure 2.5 The LAC for two plants

fundamental production technology. Synchronizing the output levels so that all are operating at their optimum is eased if they have "reserve capacity" (Koutsoyiannis 1979). This is illustrated by a flat base to the short-run cost curve, a saucer-shaped curve, and though plants may still have to be multiplied along the same principles as for matching the output of different machines (Maxcy and Silberston 1959) at least the range of optimum values facilitates this process.

At the planning stage, the long-run costs under consideration by the firm are similarly eased if there is a range of optimum values exhibited by the different plants. If plant A has a very sharp, U-shaped LAC then the alignment with plant B's LAC is more easily conducted if that for B has a wide optimum range. This is distinct from the SAC curve: a saucer-shaped LAC curve means that different sizes of plants can all be operating at the same minimum cost. The firm is not, then, tied to one size of plant for this particular process. While the company might establish one plant A and one plant B, from a planning perspective it can look to expand with varying sizes of plant B without encountering diseconomies of scale. Figure 2.5 shows this for the two plants, A and B, where LAC_A offers little in the way of alternative plant sizes while LAC_B has a wide choice of plant sizes. However, if LAC_B is still flat where the output of LAC_A is doubled, shown by point $2X_1$, for two plants of A producing at output X_1 to match the output of just one plant of B producing at output $2X_1$, plant B would result in the lowest unit cost. This introduces an additional degree of flexibility to long-term planning.

The ability to combine plants effectively, such that each is operating at minimum cost, is particularly relevant to industries involving multiple products. The plants are likely to have none or few technical commonalities, and so the only common link is in final assembly. If each manufacturing process comes under the remit of a single company, then it is the company's responsibility to align the plant outputs with each other in order that an overall MES for the company might be achieved. Finding the precise quantity of output necessary for MES can be difficult to calculate but it may be inferred when certain sizes of firms have shown significant longevity. I will now discuss the method for doing this.

2.3 Survivor analysis

As Pratten (1971) pointed out, quantifying economies of scale is fraught with difficulty since it is not possible to inflict experimental research techniques on commercial operations. Neither is it possible simply to assume that the firms that are ranked the highest in terms of output have the greater access to economies of scale since ranks may change without revealing the variation in output quantity that brought about the new rank (Hymer and Pashigian 1962). Instead, Rogers (1993) suggested that the three possible approaches to determining the relationship between plant size and production cost are econometric, engineering, and survivor analysis.

The econometric cost curve estimation appears highly quantifiable but suffers from difficulties in measuring certain costs, such as normal profit and cost of capital. The engineering approach looks at the costs of the inputs, but the cost structure can be skewed by the relative availability of particular factor inputs, thus precluding direct comparisons between plants in different locations. The approach can explain where efficiencies have occurred, but it is not predictive in showing which plant will be the most efficient. In any case, such definition of costs may not be necessary. As Frech and Ginsberg (1974) stated, it is more important to be able to understand that those involved in the activity will strive to maximize some utility function and will therefore alter the size of their organization should they observe one that is better able to maximize it. Altman (1999) asserted that it is only necessary that a firm behaves in a manner that is consistent with its survival. George Stigler (1958, 54) criticized the econometric and engineering approaches for failing to take into account the effectiveness of the inputs: "as if one were trying to measure the nutritive value of goods without knowing whether the consumers who ate them continued to live."

He proposed an alternative, the survivor principle, which took a retrospective, longitudinal view. This determined economies of scale based on the Darwinian assumption that the companies that out-lasted the competition must enjoy the lowest costs, or, in more general terms, an advantage that was contingent upon the size of their output. The theory proposed that the most efficient company size would take the largest share of the market, and by extension the most efficient company would operate with the most efficient plant size. Stigler used this technique to investigate the existence of economies of scale in the US steel industry, each plant size representing a short-run cost curve tracing out part of the long-run cost curve for the industry. The smallest and largest plants lost market share over the two decades of the study, suggesting that they operated in those parts of the LAC curve that brought decreasing returns to scale or suffered diseconomies of scale. Medium-sized plants tended to survive, suggesting that they operated in the extended flat section at the bottom of the curve.

Stigler's study showed that the survivor approach observes firms converging on an optimally sized MES, or MEPS in the case of a single plant firm, during which time they take an increasing share of the market while their smaller rivals leave the industry or are consolidated into larger groups. This implies that firms that have been in the industry for longer will hold an advantage since they have had more time in which to reach the most efficient size. Dunne and Hughes (1994) tested this amongst British companies and found that the smaller, younger ones showed the greater tendency to grow fast. This suggests that as firms converge on the optimal plant size the rate of increasing returns to scale decelerates, implying a flattening to the LAC curve where growth becomes progressively less important.

Klepper (1996) drew parallels with the product life cycle theories that once a new product has reached maturity it is the manufacturing process that comes to prominence. However, this suggests the emergence of a dominant design for the product, which Klepper considers to be an imprecise concept, particularly where market demand is diverse. Nevertheless, Klepper (2002a) has found that the larger incumbents held the advantage because the R&D costs of innovation could be allocated over a larger output. Over time Klepper states that R&D can be costlessly imitated, and it might be suggested that this would be the point at which small firms could enter with commonly held technology and so expand rapidly. The greater advantage, though, lies with entrant firms that are in some way related to the incumbents, perhaps comprising staff from those firms or being diversifications from related industries (Klepper 2002b). The Ford Motor Company, for example, was the third firm to be founded by Henry Ford and the remnants of the company immediately preceding it even metamorphosed into the Cadillac company under a different management.

The evolution of a firm is therefore better understood if the industry is taken as the unit of analysis, rather than the product, indicating an industry life cycle. Cantner *et al.* (2006) found that the firm specific factors, such as capital and initial endowments, were major determinants of a new entrant's ability to survive in the long term. Although intensity of competition reduced survivability rates, as any growth in output was directly reflected in market share, if the market as a whole was expanding then this slowed the exit rates from the industry as output could increase, or stabilize, even as market share fell.

Although survivor analysis is useful where market shares are clearly growing in a static market, as Koutsoyiannis (1979) pointed out, this intuitively attractive technique is founded on assumptions that are not always realistic, principally that firms operate under the same set of conditions; Rogers (1993) pointed specifically to the different uses made of capital and labor. Stigler (1958, 57) was not unaware of this: "Since various firms employ different kinds or qualities of resources, there will tend to develop a frequency of distribution of optimum firm sizes."

Although this suggests that survivor analysis can only result in general indications of optimal output size, there is the additional problem of technical advances, which Pratten (1971) showed will result in plants that have different short-run cost curves. Shepherd (1967) even suggested that industry entrants may employ smaller plants simply because new technology permits them to be operated efficiently at lower outputs. Over an extended period of time the LAC curve may become more a historical record of the industry diversity rather than a scale curve to aid in production planning. It could be argued that new production techniques are firm-specific, and, because they are not available for planning by rivals, they are not, therefore, part of the industry LAC. Makadok (1999) found some basis for this but the effect diminished after a period of around eight years as rivals learnt to imitate each other, referred to by Klepper (2002a) as costless imitation.

Taking this point further, in a globalized industry where the technical factors of production are available to all participants, the principles underlying plants with firm-specific advantages will soon become disseminated throughout the industry. This will have the effect of equalizing technical resources, leaving local labor resources as one that might provide an advantage that can only be imitated by moving to the same location, assuming that the labor resources are effectively unlimited. In such an industry, the short-run cost curves that comprise the industry LAC might, in practice, imply moving plants to different locations to exploit the marginal cost advantages.

Frech and Mobley (1995) noted that a plant may operate in an environment where it receives assistance from its government in some way, and it could be suggested that this would include a requirement for incoming FDI to engage in joint ventures with local firms. The motivation for government policies of this kind is that joint ventures are an effective way of transferring the advantages of the established firms to the smaller entrants. However, if the industry is global then it is likely that government support would only maintain smaller entrants at the local level, while the multinationals continue to enjoy a dominant share of the global market. Leung *et al.* (2003) have found that large foreign banks were compromised by regulations in Shanghai that favored domestic rivals, which maintained diversity and suppressed dominant market share in the city. However, in terms of total market share, which in an international environment would be global, survivor analysis would still have revealed the overall advantage that the large foreign banks held. This may also explain the finding by Gorg and Strobl (2003) that multinationals are quicker to leave local markets than their domestic rivals when the commercial environment becomes more challenging, since they have other locations within the global market. This is not an option for the government-assisted local companies which only enjoy such assistance in their domestic environment and which is often too small to offer industry-standard economies of scale.

The degree of global market share that would signify the exploitation of economies of scale, though, is difficult to determine. Shepherd (1967) cautioned that the technique is descriptive, rather than normative, and predictions based on the measured trends therefore reveal only the current MES, not necessarily the technical optimum. Growth of market share is therefore observed to be converging on an optimum that can only be inferred, the state of the industry being an indication of the current MES for the firm and, by extension, the MEPS for the plants that comprise it. Indeed, Shepherd suggests that the MEPS is the more reliable measure since the firm as a whole, particularly one of a multinational nature, is subject to a multitude of complex factors. It might be argued, for example, that in the study of foreign banks in Shanghai conducted by Leung *et al.* (2003) that local market distortions mean that output there should not be included in the foreign banks' global market share. So, while the total output of the banking firm with the largest market share may not necessarily be exploiting all the available economies of scale in the industry, it is quite valid to assume that its largest branch in a market less encumbered by local distortion is operating at the MEPS.

On this basis, the more that an industry is globally spread the more reliable the survivor analysis, since the various local anomalies will tend to equalize. Furthermore, the survival trends will be more reliable if they are well established over time, based on a shared production technology, one where progressive developments are refinements of the existing system rather than a distinct new system. This particularly applies to the automobile industry as long as it is narrowly confined to the mass-market passenger car manufacturers using the unitary all-steel welded body for their products. There are other technologies available, but it is this load-bearing body that has become the dominant feature of the volume industry, defining both the product and the production system. Since this foundation for the industry became established in the 1920s it can be assumed that the high entry costs mean that plants are only constructed after rigorous economic evaluations, based on long experience, have taken place. On this basis it would be unnecessary for a research program to investigate the minutiae of engineering or econometric costs, the evolution of the industry for the preceding decades being sufficient to establish reliable survival trends.

Chapter 3

Economies of scale in the automobile industry

Recent discussion of productivity in the automobile industry has been focused on the rise of flexible manufacturing, even suggesting that this phenomenon now takes prominence in defining the costs of production. However, flexible manufacturing does not redefine the production technology but is instead a method for more fully utilizing it. Truett and Truett (2001) assert that the presence of economies of scale in the automobile industry continues to be a significant factor for companies, recommending that the Spanish industry, for example, would find cost advantages in expanding. The expansion would not necessarily be uniform across all the production processes, since each process can have a distinct MEPS.

There are four processes that comprise the production of automobiles, as described by Rhys (1972), Wells and Rawlinson (1994), and Wells and Nieuwenhuis (2001):

1. R&D: involving the proposal for a new vehicle and development for production.
2. BIW: the pressing of steel sheets into vehicle body panels, which are then welded together to form the load-bearing vehicle body.
3. Powertrain: the fabrication of engines and transmissions using forge, foundry, and machining techniques.
4. Final assembly: painting of the BIW in the paintshop, subassembly of major components, and final assembly of the complete vehicle.

Pressing and powertrain production use specialized manufacturing processes as dictated by the nature of the product, and so are capital intensive. Pressing is the most specific to the automobile industry because it stamps out panels from sheets of steel which are subsequently welded together to form the load-bearing structural bodyshell, known in its pre-assembly unpainted state as the BIW. Assembly is a logical sequence of activities that has developed over time and tends to be more labor intensive. Vehicle design is a knowledge-based activity encompassing all the R&D necessary to bring a new model to the point of production. R&D is dependent upon human knowledge, aided by physical prototypes and, increasingly, information technology (IT).

As distinct activities, each process has its own economies of scale. Husan (1997) stated that the MES has changed as the industry has developed, notably between the years 1958 to 1975. It is tempting to agree that manufacturing has become more capital intensive but it does not necessarily follow that technical progress leads to greater economies of scale. It may instead lead to a flattening of the LAC curve for the plant, allowing the MES to be gained at lower output levels than previously, particularly with the rise of flexible manufacturing. Indeed, in reviewing the four main processes, Rhys (1999) took the view that economies of scale in panel pressing had halved to one million units per annum since his previous study (Rhys 1972), had fallen by a quarter in powertrain production to 250,000 units per annum, and yet had risen from 200,000 units to 250,000 units per annum in final assembly. Rhys (1999) also suggested a figure of 2 million units per annum to cover the R&D costs. These figures seem to contradict Husan's view that the MES increases over time, at least with respect to pressing and powertrain, though the MEPS for final assembly does appear to have risen.

The figures from Rhys (1999) would suggest an overall MES for the industry of 2 million units a year, with the manufacturing plants being multiplied as necessary to meet this common denominator. I will now employ survivor analysis in an attempt to reveal the MES for the current automobile industry.

3.1 Survivor analysis of the automobile industry

This survivor analysis of the global automobile industry takes the years 1975 to 2005 as the period for investigation, sampling the data at five-year intervals. In taking a three-decade period it is ten years longer than the study conducted by Stigler (1958). The 1975 to 2005 period commences with the coming to global prominence of the Japanese automobile industry, the later rise and consolidation of the Korean industry and, lastly, the recent emergence of the industry in China. All the production figures have been sourced from the OICA, either directly (2000–05) or indirectly (1975–95) from Ward's Automotive Year Books (1976, 1981, 1986, 1991, 1996). These published figures have originated, in turn, from the national organizations that collect the data. The most recent results have been omitted from the

Table 3.1 Top 13 automobile producing nations, 2006

Country	Production (units)	2005–06 change (%)
Japan	9,756,515	8.2
Germany	5,398,508	0.9
China	5,233,132	70.0
USA	4,366,220	1.0
South Korea	3,489,136	3.9
France	2,723,196	−12.5
Brazil	2,092,029	4.1
Spain	2,078,639	−0.9
India	1,473,000	16.5
Canada	1,389,536	2.5
UK	1,442,085	−9.7
Russia	1,177,918	10.3
Mexico	1,097,619	10.9

Source: OICA (2008b).

survivor analysis since they are often subject to revision, particularly with regards to emerging economies. Output figures from 2006 will, though, be used to illustrate the most recent changes, particularly if they are considered to come from a reliable source.

The growth in the global production of automobiles shows no sign of slowing down, although much of the most recent increase is due to the expansion of output in China. Table 3.1 shows how the top 13 automobile manufacturing countries in the world have continued to grow a year on from the survivor analysis period, with all of them producing above 1 million units a year in 2006 (unrevised figures), along with the rate of growth over 2005.

Although the countries that have dominated the automobile industry for the longest period of time, such as Germany, France, and the US, continue to have a substantial presence, it is Japan that now holds sway. It also continues to grow at a respectable rate, the 2006 output figure being an increase of 17.7 percent over the 2000 result (JAMA 2007b). Other countries have shown much faster increases in output over 2005 – 41.6 percent in the case of the Czech Republic – but it is China's combination of high output and continued high rate of growth that indicates its possible future status as the industry leader. China even overshadows India, which has also shown remarkable increases in production.

It is because of locations like China and India that global production is rising with each successive year. Figure 3.1 traces the change in total global

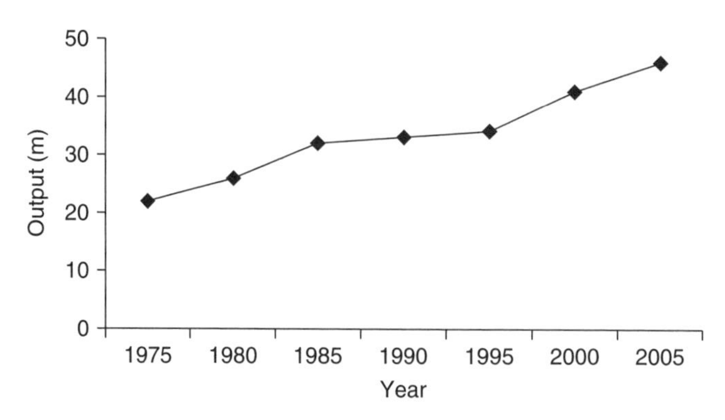

Figure 3.1 Total global automobile production, 1975–2005
Source: OICA (2008b).

production since 1975 in five-year intervals. As Cantner *et al.* (2006) have found, the growth suggests that smaller firms will be able to maintain a presence in the industry for longer than if the total global output were stable or declining, and it may even encourage successive entrants to come forward.

In order to apply survivor analysis, the total output for each year has been broken down into the contributing output figures from the individual automobile manufacturers. The range of output extended from as low as 29,000 units on one occasion, by UAZ of Russia, to over 6 million units by Toyota. The individual output results were allocated to output ranges and then the production share of the total output by those ranges was calculated. Initially, these output ranges were taken in 1 million unit groups, except for output in the zero to 1 million units which was divided into two ranges of 500,000 units each. This was in order to enable us to examine the movement of smaller companies at the lower limit of the industry. For clarity, the groups have been labeled in output ranges of whole millions (e.g. 1 million to 2 million), though in fact the upper figure in each range should strictly be one unit below that shown. The results for this are shown in Figure 3.2.

The level of aggregation of output data shown by Figure 3.2 appears too complex to reveal a single overall trend. On closer inspection, the two lowest output ranges, from zero to 0.5 million and 0.5 million to 1 million, reveal a relatively consistent decline. From 1975 to 1985 the two ranges alternate, suggesting movement of manufacturers between the two. From 1990 to 2005 both ranges are in decline in terms of production share, indicating that survival rates below 1 million units a year are deteriorating. Given

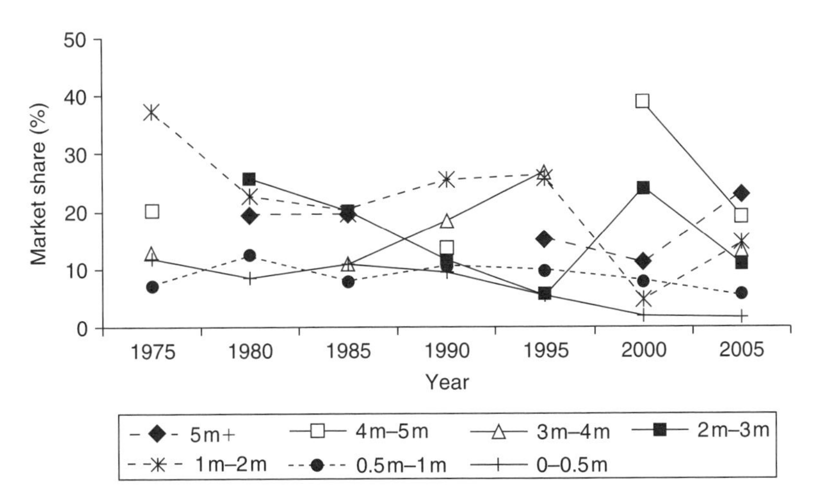

Figure 3.2 Global production share of output, seven ranges

Source: OICA (2008b).

that throughout this period the majority of manufacturers in this range are Chinese, it is demonstrated that government support through enforced joint ventures with foreign companies is not enough to sustain these companies. No Chinese manufacturer has yet broken into the 0.5 million to 1 million range. However, this also illustrates the weakness of survivor analysis in a growing market where production share may decline while output is stable or even growing. For example Chery is making excellent progress and SAIC is gradually expanding its presence as an independent manufacturer. Subaru has also consistently grown since 1975, starting the period with an output of just over 100,000 units and ending it on 500,000 units. These companies may also be joined by other illustrious companies such as Mercedes-Benz and Chrysler as their groups deconstruct.

The other output ranges in Figure 3.2 are even less clear. The 1 million to 2 million range is in general decline, though it has wide fluctuations that are mirrored in the opposing variations of the 2 million to 3 million range, also in an apparent decline. The 3 million to 4 million range emerges inconsistently but appears to share some synchronization with the 4 million to 5 million range which is even more sporadic. The range above 5 million units a year surfaces only towards the end of the period, as might be expected given the growth rate of the global production total, and too little time has elapsed for this to be identified as a significant trend.

To bring greater clarity to the data the three middle output ranges were recategorized into two, giving 1 million to 2.5 million and 2.5 million to

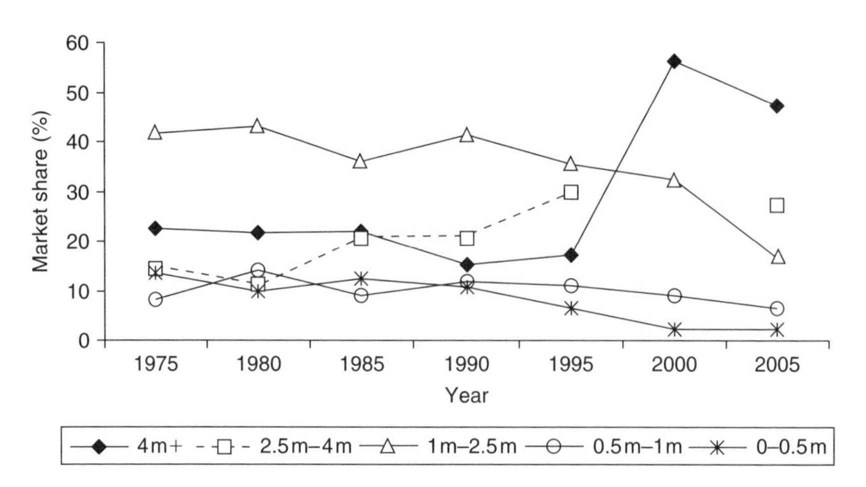

Figure 3.3 Global production share, five output ranges
Source: OICA (2008b).

4 million, and the top range reduced to 4 million units or more. The results of the five output ranges are illustrated in Figure 3.3 above.

Figure 3.3 shows the two lowest output ranges as before but it brings out the decline in the 1 million to 2.5 million unit range more clearly. The share of global production enjoyed by those manufacturers in the 2.5 million to 4 million range has been on an upward trend since 1980, though there is no data for 2000 since they appear to have moved into the range above, judging by the sudden rise by the 4m or more range. This is borne out by inspecting the production shares for individual companies for the five highest ranked companies, the industry leaders. Figure 3.4 shows this for Toyota, GM, Ford, and VAG, with data for Renault and Nissan as separate organizations until their alliance began in 1999.

Although Figure 3.4 appears to show a convergence in production share to around the 12 percent level, excluding Ford's falling share, extrapolating the trends would suggest that Toyota will continue to accumulate production share at the expense of the others, intimating a future divergence of the trends. In addition, in an expanding market even a declining production share will translate into output growth, and this is shown in Figure 3.5.

The five manufacturing groups shown in Figure 3.5 appear to be converging on each other but it is not possible to estimate what the ultimate level of output might be as long as the market continues expanding. An average of the outputs for each year could be used to illustrate a trend line over the

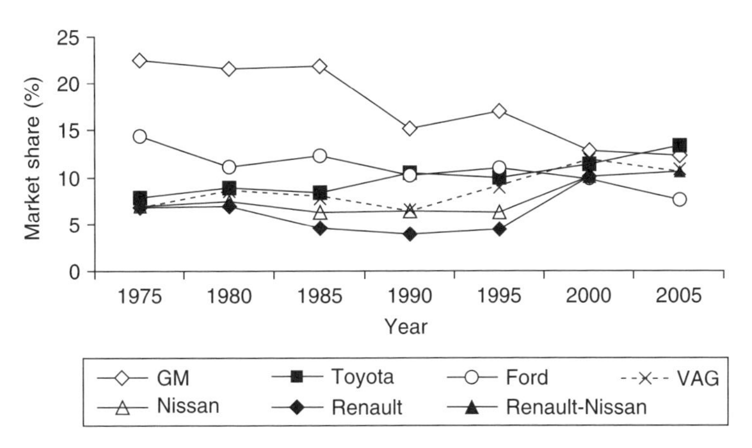

Figure 3.4 Production share for the industry leaders, 1975–2005
Source: OICA (2008b).

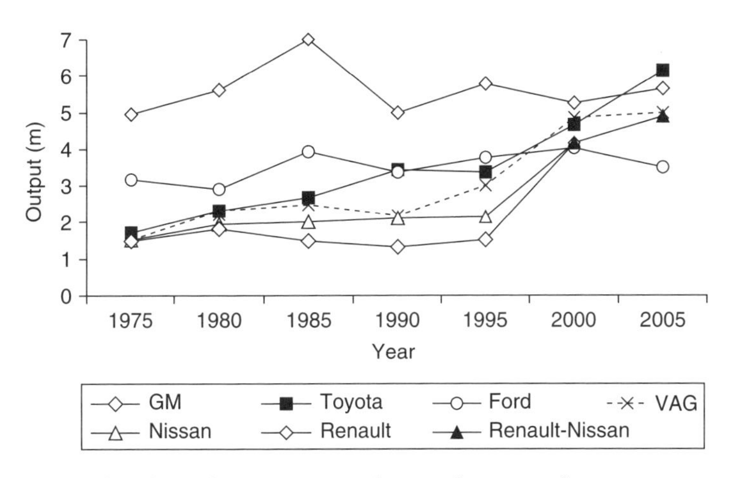

Figure 3.5 Global production output for top five manufacturers, 1975–2005
Source: OICA (2008b).

period for these manufacturers, and, although this would give an output figure for 2005 of a little over 5 million units, the variance on either side of this is over 1 million units. Furthermore, the gradient of the average would be clearly positive and continue to be so as long as the market continues to expand. According to forecasts by the automotive consultancy IHS Global Insight, the world production for automobiles will expand another 28 percent to over 60 million vehicles by 2010, much of this due to further

Table 3.2 Top 20 automobile manufacturers, 2005

Rank	Company	Output (m)
1	Toyota	6.2
2	GM	5.7
3	VAG	5.0
4	Renault	4.9
5	Ford	3.5
6	Honda	3.3
7	PSA	3.0
8	Hyundai	2.7
9	DC	2.0
10	Suzuki	1.7
11	Fiat	1.5
12	BMW	1.3
13	Mazda	1.1
14	Mitsubishi	1.0
15	Daihatsu	0.8
16	AvtoVAZ	0.7
17	Fuji (Subaru)	0.5
18	Chery	0.2
19	Tata	0.2
20	SAIC	0.1

Source: OICA (2008b).

increases in China, where production will be over 12 million vehicles that year. If Toyota maintains the trend in production share growth shown in Figure 3.4 since 2000 then a production share of 15 percent in 2010 would give it a production output of 9 million vehicles. This is corroborated by Toyota itself which has declared as its target a 15 percent share of the global car market by 2010 (Toyota 2007a). A further difficulty is that as the market expands manufacturers release more models. Toyota, for example, lists 53 models in production in Japan alone (Toyota 2007b), although the number of vehicle platforms is much less.

Survivor analysis seems to show that output ranges of less than 2.5 million units a year have a poor record of longevity, either because manufacturers move into a higher range or drop down into a lower one. Table 3.2 shows the top 20 manufacturers for 2005, of which only eight manufacturers were above this supposed threshold. Of those that are under the threshold there are some that can call on support from equity holders, such as Suzuki (Toyota), Daihatsu (Toyota), and Mazda (Ford). This may not in itself be relevant since equity holders seek a return on their investment and so it would be expected that these companies should strive to exploit economies of scale just like independent companies. Alternatively, BMW has no external support but is able to practice cost recovery due to its premium brand image.

Although the survivor analysis has been effective in identifying higher output ranges comprising companies that might be considered the industry leaders, it has not conclusively demonstrated that the remaining output ranges, those with falling output shares, comprise manufacturers that are in danger of being eliminated from the industry. Indeed, the growth of the largest manufacturers seems to be a fairly recent phenomenon, while the smaller companies, those below 1 million units a year, comprise new entrants together with some of the oldest names in the industry. It should also be noted that in 2007 this range was joined by Mercedes-Benz due to its demerger from Chrysler. Since the MES is, by definition, the minimum size that captures all the available economies of scale, the survivor analysis seems to indicate three conclusions:

1. In the very lowest output range the firms are mostly new entrants to the industry.
2. In the 0.5m to 1m output range there are firms that are long-term survivors in the industry.
3. Above an output of 2.5m, firms have an additional advantage that grows in proportion to size.

To investigate the MES more closely, Shepherd (1967) advised the analysis of individual plants to uncover the MEPS first, from which the MES could be derived. This assumes that all companies will attempt to achieve the MEPS, whatever their overall size, and as they expand they will do so with plants that exhibit the same cost advantages. Thus, companies will progress towards the MES in units of MEPS. The following section will evaluate the MEPS taken from the dominant and longest surviving companies in the industry for each of the four processes that make up automobile manufacturing.

3.2 From plant size to economies of scale

Of all the corporate functions that comprise automobile manufacturing, the most characteristic and restrictive is the manner in which the steel bodies are formed. From the time the automobile was born in 1876 in Baden-Wuerttemberg – where Nikolaus Otto produced the first high-revving lightweight internal combustion engine (Maxton and Wormald

1995) – until 1915, car production was a craft activity. This depended on the skills of carpenters to construct and fit an unstressed timber-frame body to a metal chassis. Even in 1920, 12 years after the introduction of the Model T, car bodies were still 85 percent wooden. Six years later car bodies were 70 percent steel.

The production revolution took place in 1915 when Dodge lifted output from 5,000 to 50,000 due to the production innovation of Edward G. Budd and his Austrian partner, Joseph Ledwinka (ibid.). The process has been termed "the Budd Paradigm" by Nieuwenhuis and Wells (1997) because it is one that distinguishes automobile production from all other industrial activity. It encompasses the pressing of steel sheets into body panels and the subsequent welding together into complete load-bearing car bodies, which give the final vehicle its shape and practical purpose. As befits a paradigm, "Buddism" is restrictive and self-referential since it refers only to the pressing and welding of steel bodies. It is not possible to substitute synthetic materials such as glass reinforced plastic (GRP), and even alternative metals, such as aluminum, can only be used with significant modifications. As a defined technology it also predetermines the available economies of scale. The following subsection will demonstrate the nature of this paradigm.

The Budd Paradigm

The paradigm has its roots in the New England clock and lock industry of 1860 (Nieuwenhuis and Wells 1997, 2007). To make the new technology fit the automotive industry Budd and Ledwinka had to redefine the basic elements of the automobile: the BIW was formed from the cowl, roof, side, and rear panels onto which body panels could be hung. The basic method of forming the panels has not changed since that time: sheets of steel are inserted into high precision dies mounted on presses and the panel shapes stamped out. Those panels that are being attached together are held in place by a jig for welding. The resultant BIW forms the basic structure of the vehicle and can be sent forward for painting, powertrain installation, and fitting out in the final assembly function.

The BIW process is capital intensive and highly amenable to automation. Indeed, the costs of the huge step-up in volume that Dodge were forced to undertake eventually led them to look to Chrysler for financial rescue (DaimlerChrysler 2004). The fixed body structure of the BIW placed physical limitations on the finished model, thus ending the craft era capability of customizing each model to the customer's taste. The Budd Paradigm is therefore central to the mass production of standardized automobiles. This was further enhanced by another major advantage of using steel alone, which was that enamel paint could be used and baked at 450°F (230°C), reducing paint drying times from between three and eight weeks to one or two hours (ibid.).

In the early 1920s, Citroën used the same technology to raise its production from 30 or 50 per day up to 400 to 500 per day. In 1925, the technology reached the UK when Pressed Steel was set up by Budd and William Morris. The Budd Paradigm was thus the technological driver that reduced the price of the final product and so created the mass market (Church 1994). Dodge continued to develop it with the 1928 Victory Six as a "monopiece" construction of inner and outer panels, followed by the Chrysler Airflow and then the Lincoln Zephyr. Budd perfected the process with the Nash 600 monocoque, which weighed around 600 pounds less than its predecessor and 300 pounds less than the body proposed by Briggs (DaimlerChrysler 2004). At this point the three main advantages of steel-body production were in place. These were:

1. Ease of manufacturing: production automation and faster paint drying times, which are suited to high-volume, low-cost production.
2. Body strength: obviates the need for separate chassis and improves passenger safety.
3. Weight reduction: improves vehicle performance and fuel economy.

The Budd Paradigm also comes at a high capital cost. Nieuwenhuis and Wells (2002) found that the typical investment required for such a plant in the early twenty-first century totaled up to £515 million (US$774 million). They estimated that the press shop cost around £100 million (US$150 million) to establish, with an additional investment of between £20 million (US$30 million) and £65 million (US$98 million) for the press tools for each model. The BIW fabrication process could cost anywhere between £50 million (US$75 million) and £100 million (US$150 million), with the paint shop adding another £200 million (US$300 million) to £250 million (US$375 million).

Fortunately, for the purposes of analysis, it is because of this technological basis to the Budd Paradigm processes that they are more amenable to quantification of the MEPS. This will be discussed in the next subsection.

Economies of scale in body pressings

The initial part of the BIW process is the shaping of the steel panels. For panel stamping alone Rhys (1972, 289) stated that the maximum usage that

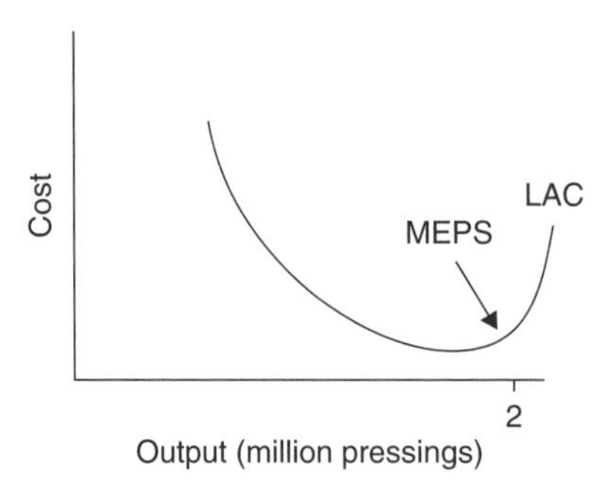

Figure 3.6 Theoretical economies of scale for panel stamping

can be expected from a press results in 2 million panels a year, since body presses appear to have an optimum speed of 2 million stampings a year. The dies can last for 7 million stampings, so the optimum number of pressings is 2 million a year over 3.5 years. Since panel production is necessarily capital intensive, the presses cannot be replaced by labor except at the very lowest output levels of craft production, suggesting a single plant size with a short-run minimum average cost equal to the long-run minimum average cost for the industry. Figure 3.6 shows this with increasing returns as production rises to the 2 million mark, but a rapid onset of decreasing returns as the machinery is pushed beyond its design limits.

However, BIW is also more than simply the stamping of panels, it also involves welding in jigs and paint preparation processes. The MEPS for the entire BIW process will therefore differ from that of panel stamping alone. Furthermore, BIW production tends to use flexible manufacturing to feed multiple models directly into the final assembly plant, more so than powertrain which can serve multiple models assembled in different plants. It is therefore difficult to use survivor analysis of the main companies in the industry to evaluate BIW in isolation from final assembly. Nevertheless, it is unlikely that BIW, being almost as highly capital intensive as panel stamping, would be allowed to operate with substantial cost disadvantages simply to suit final assembly. Taking this into consideration, it can be suggested that where BIW and final assembly operate together then a sense of the economies of scale for BIW can be inferred if necessary.

Rhys (1999) bases the 2 million output figure for panel pressing on the technical capability of the machinery. There is little empirical evidence of this occurring, for example Ford's Dearborn stamping plant presses panels for both the F-150 and Mustang sports cars with a combined production in excess of 740,000 in 2005 (*Automotive News* 2006a). The Harbour Report

2007 assesses productivity for production plants across North America (*Reliable Plant* 2007). For 2006, it concluded that the most productive for stamping was the Honda plant in Marysville, whose total production capacity amounted to 440,000 units per year (Honda 2006). The Toyota plant in Georgetown was ranked second in the report for productivity with a declared production capacity of 500,000 units a year (JAMA 2005). This suggests that the MEPS for BIW is around 500,000 units a year.

To explore the shape of the LAC curve it is necessary to investigate the viability of lower output rates. The UK Toyota plant at Burnaston had output in 2005 nearly half that of the Georgetown plant, at 264,000 units (SMMT 2006b). In Australia, the Ford Geelong plant recorded total BIW production of 101,014 in 2005, with capacity set at 120,000 units a year (*Automotive News* 2006a). Due to the location of the plant the output probably represents the priority of local assembly over economies of scale, partly due to reasons of tariff barriers. Silberston (1972, 377) stated that, in such a case, scale has a relative value: "If the market is too small to contain even one plant of the minimum optimum scale, then it follows that any plant set up to produce for that market cannot be as efficient as it is possible to be, because its scale will be too small."

The difference between US, UK, and Australian output levels is explained by the costs of transporting BIW to Australia and also the tariff structure that adds 10 percent to imported vehicles. This makes it cheaper overall to produce BIW at Geelong, even at this output level. Much below this and final assembly is usually fed by imports of BIW in complete knock down (CKD) kit form to serve the local market. For example, from 2000 the Opel Corsa was assembled from CKD kits in China at the rate of 22,000 units a year and sold as the Buick Sail (Sieradski 2002). This suggests that a stamping plant at this level of output would introduce cost disadvantages in excess of the costs of transporting CKD kits from larger plants elsewhere. In view of the various levels of output shown by different plants it would appear that the LAC curve for BIW production has a shallow gradient to the increasing returns to scale side. Based on the Harbour Report, the MEPS is around 500,000 units a year (e.g. at the Georgetown plant), with a shallow gradient from 250,000 a year (e.g. at the Burnaston plant), and the onset of diseconomies of scale beyond 750,000 units a year (e.g. at Dearborn Stamping). The LAC curve for BIW is illustrated in Figure 3.7.

In Figure 3.7 the output for the Geelong plant is just above the 100,000 level, at the point where the increasing returns to scale dictate that the curve begins to take on a shallow slope. The Dearborn plant would be near the 750,000 output level, beyond the MEPS, but the slope of the curve is shallow enough not to incur pernicious cost disadvantages. Between

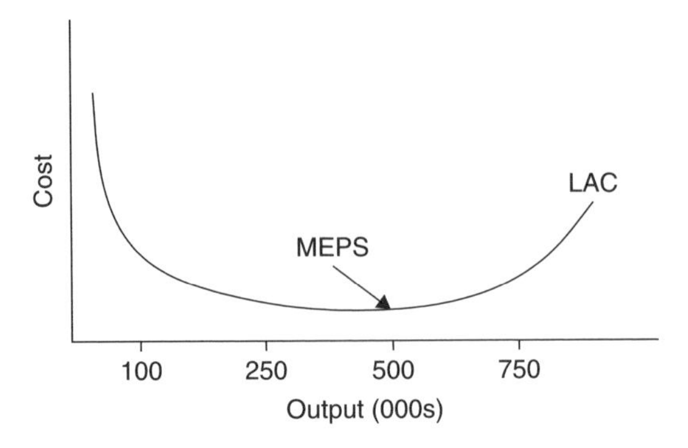

Figure 3.7 Economies of scale for BIW production

the two would be Honda's Marysville plant at 440,000 units a year and Toyota's Georgetown plant at 500,000 units a year, judged by the Harbour Report to be the most efficient amongst the competition and therefore representing the MEPS.

Economies of scale in powertrain

Powertrain production uses processes that are quite distinct from the pressing of shapes into sheet steel that characterizes the BIW function. Engines and transmissions are made up of intricately shaped metals that are designed to perform under highly stressful conditions to the extent that parts stamped from sheets of material would not be robust enough. Powertrain parts are therefore fabricated by either casting or forging.

Casting involves producing the basic shape of the component by pouring liquid metal into a mold and then machining away the surplus material to arrive at the final, desired specification. This is commonly used for engine blocks, cylinder heads, and gearbox casings. Forging produces components by hammering them into shape, usually when the material is hot enough to be malleable. The desired shape is arrived at by plastic deformation of the material, which also introduces a degree of work hardening to increase the component's mechanical strength. This is applicable to components such as crankshafts, camshafts, and piston rods. Although the two processes are technically unrelated there is a degree of substitutability. For example, gears can be cast as blanks, rather than forged, before the precise shape is machined into them (*Gear Product News* 2006).

Bhaskar (1979) suggested that since more than one engine variant would be required for the complete model range the powertrain plant would need to have capacity for between 500,000 and 1 million units a year. This

Table 3.3 Japanese powertrain capacity in the US, 2004

Plant	Output	Capacity
Honda engines, Ohio	1,079,564	1,160,000
Honda transmissions, Ohio	1,003,289	1,000,000
Nissan engines, Tennessee	728,986	950,000
Toyota engines, West Virginia	456,231	540,000
Toyota transmissions, West Virginia	389,859	360,000
		600,000 (2007 projected)

Source: JAMA (2005).

appears to be the current requirement in 2006, Ford stating a target output for its British plants in Bridgend and Dagenham of 1 million engines a year by 2009 (Ford 2006). The Bridgend plant produces three types of engine on two lines: four cylinder engines on one and in-line sixes and V8s sharing the other. Toyota Deeside produced just fewer than 500,000 engines in 2005, though more than half were exported in component form to other Toyota plants for final powertrain assembly (SMMT 2006b). In the US, the market is large enough to invite plant investments that explore the economies of scale further. Table 3.3 samples data published by JAMA to show this for the greenfield investments by Japanese firms.

Table 3.3 indicates the range of outputs where economies of scale might apply: from around 500,000 for Toyota up to around 1 million for Honda and Nissan. Since Toyota powertrain capacity is the same for the UK and the US, this implies that Toyota has settled on a size of plant that exploits the available economies of scale. Honda, though, shows that a scale of 1 million is possible and Ford demonstrates that worldwide distribution creates the demand conditions where output of this level can be planned for, even when manufacturing takes place in a market as small as the UK. Figure 3.8 shows the LAC for powertrain production where the SATC for Toyota plants would have their minimum cost tangential with the LAC curve at 500,000 units a year, while Ford and Honda plants would have it at 1 million units a year. This LAC curve shows that Toyota is not suffering significant cost disadvantages with such an output, despite its plant being half the size of that of the competition, and so the LAC curve has a shallow slope until the MEPS at 1 million units.

Final assembly processes

Final assembly has come to symbolize the scale of car production more than any other process yet, in practice, it is an uncomplicated procedure

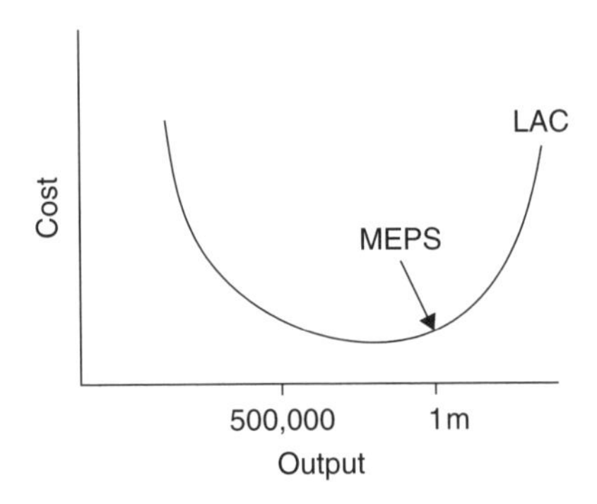

Figure 3.8 Economies of scale for powertrain production

that was arrived at through exploration of work practices rather than the development of a core technology. Henry Ford introduced many of the flow production line techniques to the automobile industry and, as a logical process, it has become the standard assembly method for the world's automobile industry, often referred to as "Fordism" (Hounshell 1984). Fordism, though, was not a single idea but an approach that integrated and refined production processes while restructuring employment practices. Dankbaar (1992, 3) stated that Fordism was more than simply mass production: for it extended into mass consumption – since higher wages paid for production translate into higher disposable incomes for consumption.

The principle of the moving assembly line is quite simple, the product being moved along stations to be progressively assembled by stationary employees working from parts delivered to them. It is not a novel concept, its use predating the automobile industry itself (Hounshell 1984; Lewchuck 1987). Even later developments such as the Toyota Production System (TPS) and lean production amount merely to detailed refinements of the overall process rather than a fundamental change in the underlying principles. It is therefore relatively straightforward for final assembly plants to produce more than one model, and in fact there are few single model plants in operation.

Economies of scale in final assembly

For reasons of competitive advantage, the quantitative costs associated with any part of production are not usually published, so econometric and engineering evaluations of scale are unsuitable; but by using survivor analysis the plant level strategies for production can be analyzed. Some additional information comes from the interviews conducted by this research.

Table 3.4 Japanese assembly operations in the US, 2004

Plant	Output	Capacity	Utilization	Employees	Investment $m
Mitsubishi, Illinois	112,984	240,000	47%	1,786	1,514
Nissan, Mississippi	267,350	400,000	67%	4,100	1,430
Toyota, Indiana	374,048	300,000	125%	4,659	2,600

Source: JAMA (2005).

The results shown in Table 3.4 for capacity utilization, as sampled from JAMA sources, reveal the problems in making quantitative calculations. Toyota has the highest output yet seemingly lower capacity than Nissan, while Mitsubishi is much weaker with less than 50 percent capacity utilization. The problem here is that the capacity statement does not take account of the installation investment. Toyota has by far the highest investment in its plant, so higher output is only to be expected, implying that the company has installed for the higher output. Nissan may have the physical plant area to offer higher capacity but the investment suggests that the installation is for lower production. Not only is Nissan's investment markedly lower than Toyota's but it also has a higher labor content per unit, indicating that the quoted capacity may be higher than the installation could achieve in practice. Mitsubishi's investment is slightly higher than Nissan's, but it had to reconfigure its installation to align with lower US sales. The company reduced the Illinois plant to a single working shift with an output capacity of 140,000 units a year for three car models and one SUV (*Automotive News* 2004b). This accords with Truett and Truett (2003) who have found evidence that final assembly was competitive when output was at least above the range of 150,000 to 200,000 units a year. I have learned from the Japan Automobile Manufacturers Association (JAMA) that assembly needs to be at least 200,000 units a year to exploit technical efficiencies.

Mitsubishi's output might indicate that 140,000 is the lowest limit for the MEPS where the local market is large enough to permit pursuit of industry-standard economies of scale, in contrast to Australian plants where they are constrained by the limited local market. However, Toyota did not reveal a marked tendency towards the larger size of powertrain plant (see p. 55), so in final assembly, unless there are constant returns to scale regarding Toyota's output, it is likely that there will be a shallow downward slope from output of 150,000 units a year to the latest data on the MEPS at around 350,000 units a year. It is also reasonable to expect

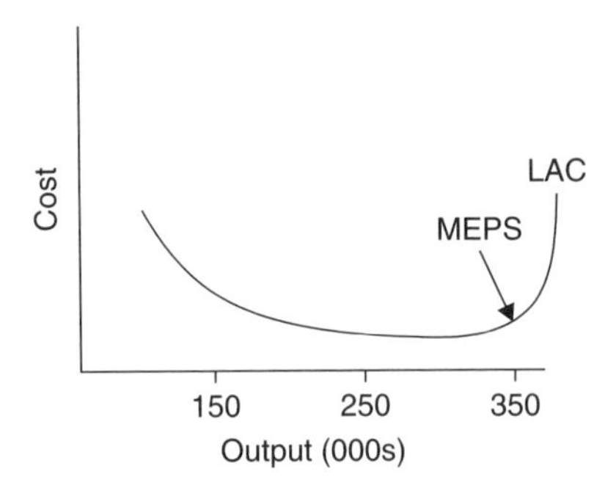

Figure 3.9 Economies of scale for final assembly

that the gradient of the curve will be particularly shallow after the figure of 250,000 units a year, the approximate capacity of the Toyota plant at Burnaston in the UK (SMMT 2006b). This is shown in Figure 3.9.

Due to the greater use of labor for smaller plants it is reasonable to expect that on the increasing returns to scale side of the LAC curve the slope will be relatively shallow. This is because it is relatively easy to design a production line with varying quantities of labor, but only up to a certain point. For example, my research has found that at Bentley output of 3,500 units per year was viable with a workforce of around 1,000, but this was for a product for which a cost recovery pricing strategy was possible. For mass-market production, less than 150,000 units a year would appear almost unviable; and beyond 350,000 units a year there will be an overcrowding effect within the labor force. The production capital will also reach its design limit, leading to a steep rise in diseconomies of scale. Since flexible production techniques mean that final assembly plants can produce a number of models, this aids in maintaining production rates; for example, the Bentley line currently assembles three variants of the GT. However, if net output capacity exceeds anticipated quantities of market demand then actual output will necessarily suffer: Hyundai's US plant was forced to suspend production for two weeks due to the rise in production of the Santa Fe sports utility vehicle (SUV) being insufficient to offset the drop in production of the Sonata saloon car (*Automotive News* 2007c).

The problem of scale in R&D

R&D, which includes all the design tasks necessary to bring a product to the point of production, is the source of long-term sustainability for a firm. The overall LAC curve for an industry implies that new products will become available as necessary to fulfill the chosen capacity of the plant. Some industries are more R&D dependent than others, the top three highest spending being technological hardware, pharmaceuticals, and the automotive industry, with the vast majority of the spending being conducted by companies in the US, Japan, Germany, France, and the UK (DTI 2006).

Although absolute levels of R&D spending reveal the monetary amounts that are necessary to compete in an industry, they do not show the strategic importance that the activity has within a company. For this reason, R&D intensity is usually measured as a percentage of sales (Rosegger 1986). However, this does not show the relative productivity of R&D, i.e. the efficiency in generating innovations. For this reason R&D should also be related to the number of employees or the number of product lines in order to express R&D efficiency; yet such data is usually considered commercially sensitive.

The data that is widely available is that of total expenditure since it often has to be declared for regulatory reasons. Even before investigating the data it is obvious that the higher levels of R&D spending will be more affordable for larger companies. There may then be economies of scope as larger firms can spread capabilities across a wider product range and diversification of risk as they absorb instances of failure more easily. Larger budgets may also provide protection from smaller entrants to the industry due to the high barriers to entry created by the absolute R&D spending levels (Thompson and Formby 1993). Despite this, larger R&D budgets and economies of scope in the product range do not mean that the firms are more efficient and can take firms into diseconomies of scale in the R&D function. Parkin *et al.* (1997) argued that there may even be a strategic pressure to increase R&D spending as companies compete to bring new products to market. This suggests that, even if the industry might be viable without R&D expenditure, if one company invests then the others are compelled to match it in order to maintain their strategic position. The alternative is the risk that they might suffer disproportionately heavy financial losses should they miss out on a crucial innovation. It is the asymmetry of risk that induces the companies to intensify their R&D spending.

Although there may be sound strategic reasons for operating with large R&D teams this does not negate the existence of an MEPS for R&D, one that could be met by using smaller teams. Cooper (1964) puts forward three possible reasons to explain the greater efficiency of smaller R&D departments:

1. Comparatively higher technical capability in the engineers.
2. Greater personal attention to cost control.
3. Improved communication and coordination.

The research by Cooper (1964) covered firms in a variety of industries. One executive, who was responsible for a large department of over 1,200 engineers, reported that this size of department was advantageous when dealing with large projects for which the personnel could be divided into specialist teams. In so doing, this seemed only to provide further evidence that smaller teams were more effective, although the executive also felt that his department held a breadth of experience that brought economies of scope to larger projects. A distinction should therefore be drawn between R&D functions that are efficient, in terms of achieving economies of scale, and those that are effective in achieving their design briefs, thus benefiting from economies of scope. In order to achieve higher economies of scope in the product range greater product differentiation is necessary, entailing greater use of R&D resources. For R&D to be effective in putting forward diverse products the function may have to grow even beyond its efficient economic scale.

The minimum size of a team depends on the nature of the task and the divisibility of the necessary inputs. It might be possible for a research team to be as small as two: one conducting the experiments and the other writing the reports. For larger projects the research team will have to expand, although coordination difficulties may introduce diseconomies of scale. Since the activity is not machine paced there will be no sudden rise in the cost curve as the equipment reaches its physical operating limits. It is likely, therefore, that the cost curve will have shallow gradients. Cooper (1964) investigated large firms in various industries with different sizes of R&D teams, ranging from 150 engineers up to 1,200 for a large department. Figure 3.10 illustrates these findings.

Figure 3.10 shows the MEPS when the R&D team is relatively small, around 150 engineers, with costs rising gradually, so that a team of 1,200

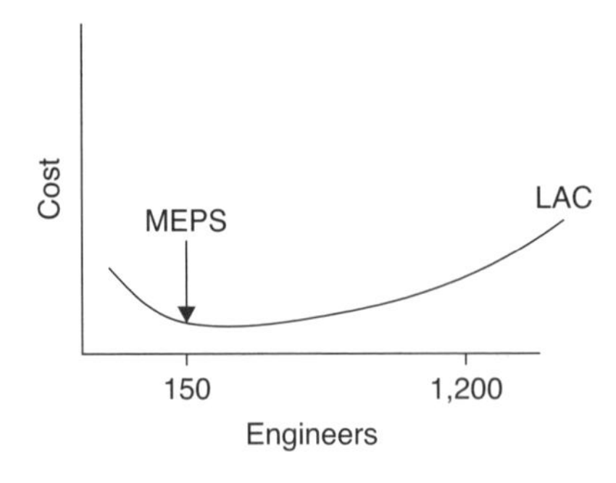

Figure 3.10 LAC curve for R&D, various industries

Source: Cooper (1964).

engineers has a cost disadvantage, although it may be more effective if the task is large and there is a time constraint. At some point beyond this level, costs will rise to a degree that not even the effectiveness of the team is enough to counter the higher costs and it will be necessary to divide the task between separate teams, perhaps by engaging different companies. In the next subsection I will evaluate this in a specific industry, that of automobile manufacturing.

Economies of scale in automobile R&D

R&D is the one function in an automobile company that eludes decisive estimation of efficiency. It has a complex interaction with the other functions, as well as market demand, such that the requirement for additional models may push the R&D function into diseconomies of scale. The function includes product concept generation, product planning, product engineering, and production engineering (Clark *et al.* 1987). Development costs are mostly determined in advance so, in theory, the higher the production the lower the proportion that development costs will take of the total cost. Dunnett (1980) found that production figures supporting R&D in the late 1970s ranged from "significant" to 5 million units a year. Rhys (1999) puts the figure at 2 million units a year.

In 2006 there were a number of manufacturers producing less than 1 million units a year in total but which maintained an R&D operation. The smallest of these was probably Proton with production of 144,395 units, but with substantial government assistance the company cannot be said to be independent. The Chinese entrant Chery had an output of 308,425 units for the year and, courtesy of input from numerous external consultancies, maintained control of its own R&D. Again, though, this was also with significant political assistance. Subaru is owned by Fuji Heavy Industries, with Toyota as a minority partner, and has output of a little less than 600,000 units a year with its own R&D capability. The smallest standalone producer of passenger vehicles is BMW, all other smaller producers being members of parent groups, and it achieved output of 1.4 million vehicles in 2006.

The problem with any output figure is that it is dependent on the demands of the market and the length of the model life cycle. One plan has suggested that the MG-TF sports car would be profitable with sales of 20,000 a year (*Autocar* 2005), and with a similar ten-year life cycle to the Mazda MX5 this suggests a total of 200,000 for this model. Of course, with a life cycle of this length it is unlikely that there would be a dedicated R&D team

Table 3.5 Global R&D expenditure, June 2005–June 2006

Company	R&D expenditure (£m)	R&D expenditure (% of sales)	R&D expenditure/employee (£)
Ford	4659.8 (US$8480.8)	4.5	15,500 (US$28,200)
Toyota	3726.8 (US$6782.8)	4.1	14,000 (US$25,500)
VW	2799.9 (US$5095.8)	4.3	8,600 (US$15,700)
Honda	2308.4 (US$4201.3)	5.4	16,700 (US$30,400)
BMW	2140.3 (US$3895.3)	6.7	20,700 (US$37,700)
Subaru	256.8 (US$467.4)	3.6	9,500 (US$17,300)

Source: DTI (2006).

working on successive sports cars. Since design teams move from project to project it is preferable to consider the R&D spending over the entire model range and relate it to sales. Table 3.5 shows this for a selection of global manufacturers for the period June 2005–June 2006 using data from the Department of Trade and Industry (DTI) in order to make a consistent approach to the figures.

It is no surprise that the range in R&D expenditure is wide, and this may be explained by the breadth of the model range. This is particularly the case for Ford which has a large number of light trucks which are classed, and used, as passenger vehicles. Large firms also tend to invest in long-term projects, for example in the case of the automobile industry this would include fuel-cell research. Indeed, interviews conducted with Honda for this book found that the company readily promoted its developments not only of automobile related technology, but also robots and jet engines. Although the company claimed that these projects were related to mobility there was no credible connection to the automobile industry. Other companies were also engaged in research that was beyond the limits of automobile production and might have been indicative of their search for a new manufacturing paradigm. To demonstrate how these additional projects inflate overall R&D expenditure, R&D intensity is usually found by relating it to the number of sales it helps to capture, shown by expenditure as a percentage of sales income. This procedure shows that the companies illustrated in Table 3.5 are quite tightly grouped, BMW being the heaviest investor and Subaru spending the least for its level of sales. It is possible that Subaru receives a small degree of informal R&D support from Toyota that has not been accounted for.

Table 3.5 follows the DTI surveys in showing R&D commitment by relating total R&D to the entire workforce. This indicates the relative R&D for the different companies by showing the level of expenditure for each worker. For example, although Toyota's R&D expenditure is 20 percent lower than Ford's it is only spending 10 percent less per employee, implying that it is making a greater R&D commitment. BMW makes by far the biggest commitment, while VW spends a remarkably low £8,600 (US$15,700) per employee.

The discovery that VW is a low R&D spender per employee is difficult to reconcile with its status as a high spender of R&D overall and its leading position in the retail market. Assuming that the purpose of R&D activity is comparable throughout the industry then, by using Toyota as a benchmark, due to its known efficiency in other functions, it might be reasonable to conclude that higher spending per employee denotes a greater efficiency in its investment. Alternatively, it might be the case that higher efficiency means that VW is able to engage in R&D programs with a lower expenditure for each R&D engineer. However, VW does have a larger workforce, some 22 percent higher than Toyota (DTI 2006) and it is likely that this is weighted towards its production operations in emerging regions. Rates of R&D spending per employee are therefore not a reliable measure of R&D efficiency if it includes employees engaged in other activities.

Companies rarely publish figures for total R&D staff, although it may be possible to identify the workforce for specific R&D centers, useful if they have the capability to take a full new model program from proposal to production. Such centers in the US range from 1,300 personnel at Honda to 800 at Toyota, and even as low as 129 at Mazda. Similarly, in the UK, Lotus Engineering, for example, claims to be able to deliver complete vehicle engineering proposals using 1,000 engineers, less than Honda employs in the US (Lotus 2006). I found from interviews that Aston Martin had an R&D strength of 170 personnel and Bentley of 600, both claiming that this gave them design authority with detailed assistance on request available from their parent companies. Extrapolation from these figures suggests that a basic model range of around three distinct models might indeed be supported by an R&D strength of around 1,200 personnel.

The purpose of R&D is to propose new models, so a further measure of R&D effectiveness would be to relate expenditure to the number of models. However, there are difficulties in simply counting the number of models that each manufacturer claims to produce. Many products share platforms, but design differences allow them to attack different market niches. On rare occasions one product range may use model names to differentiate trim levels or brands with very little real difference in design; more commonly one model may be sold under different names for different markets (e.g. Mitsubishi Pajero/Shogun). There will also be models that are designed for foreign markets only, such as Honda's joint venture involvement in the China-only Guangzhou Honda model to be released in 2010.

Table 3.6 R&D expenditure per model, June 2005–June 2006

Company	Models	R&D expenditure per model £m ($m)
Ford	59	79.0 (143.8)
Toyota	81	46.0 (83.7)
VW	44	63.6 (115.8)
Honda	29	79.6 (144.9)
BMW	11	194.6 (354.2)
Subaru	9	28.5 (51.9)

Source: DTI (2006).

Table 3.6 counts the number of models sold according to company reports and websites, although it may undercount where models exist only as part of short production runs or to serve obscure markets. The R&D per model has been calculated from the DTI figures shown in Table 3.5.

Due to the various problems with precisely defining and counting models Table 3.6 can only be an estimate of the R&D spending per model, but it does give some indication of how the company strategies relate to each other. It should be noted that Table 3.6 does not show the actual expenditure for particular models since in practice there will be one core model and then a series of variants spun off from it. The figure shown is the average across all those core models and variants. Subaru spends the least per model, with nine models based on four platforms, one of which is a light Kei car platform with three variants. From Table 3.6 it would seem that it would cost Toyota, on average, £46.0 million (US$83.7 million) to put a new model onto the market, whereas it would cost BMW £194.6 million (US$354.2 million). Since Toyota enjoys far higher total sales than BMW (see Table 3.2) it is reasonable to conclude that a low R&D spend per model is an acceptable analogy for an efficient R&D operation. As in a previous estimate of R&D efficiency which related spending to sales, Subaru again shows itself to be the most efficient.

This research estimates that the MEPS in R&D lies somewhere close to that of Subaru, with the proviso that it might be receiving some assistance from Toyota. Using production data from 2006, the year of the DTI survey used, Subaru appears to show the following conditions for MEPS in R&D to be met: output of 600,000 units per year and expenditure of £256 million (US$471 million) per year over four platforms and nine models. Since the Kei cars are peculiar to Japan it could be suggested that Subaru has a core range of three platforms and five models in order for it to be comparable as a conventional automobile company. The company is not entirely self-sustaining since it can call on resources from affiliated companies. The value of these relationships is difficult to quantify but it would probably offset the

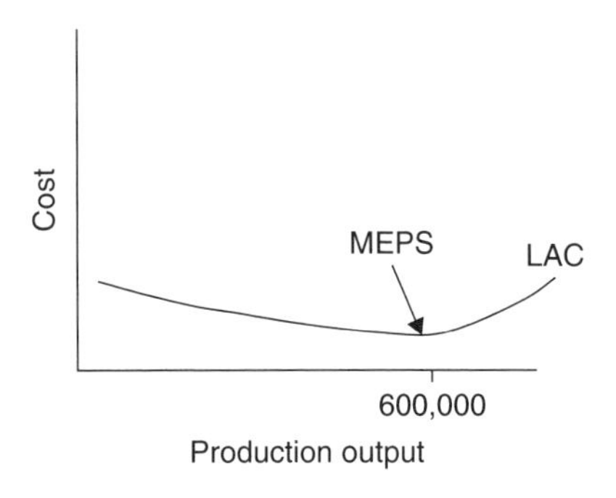

Figure 3.11 Economies of scale for R&D

subtraction of the Kei car line from the production and R&D expenditure data. This suggests that an automobile company with a similar range of conventional products (i.e. three platforms, five models) would need an output of around 600,000 units a year and an R&D investment of around £250 million (US$460 million) a year. Data from other empirical sources, including interviews conducted for this book, suggest that the R&D personnel should number around 1,200 engineers.

Figure 3.11 illustrates a possible LAC curve for R&D, based on the level of production output that it both serves and is served by. Since R&D is dependent upon human assets rather than being paced by machinery, the shape of the curve is much shallower than it would be for capital intensive processes like BIW. Up to the optimum of around 600,000 units a year there will be some increasing returns to scale as teamwork effects and economies of scope bring additional benefits. Thereafter, diminishing returns to scale will set in as a crowding-out effect appears. For production output that is higher than 1 million units a year, it is feasible that multiple R&D centers will be employed. It should be noted that the minimum point appears to have shifted to the right compared to that shown in Figure 3.10. This is because an automobile is a complex product and so the R&D team will be divided into smaller groups depending on each aspect of the project.

Having determined the LAC curves for the four functions at the core of the automobile firm, these will now be combined in order to reveal the MES for an automobile manufacturer.

Combining the LAC curves

From the estimates of economies of scale, both published in the literature and provided by the research interviews, for the three main automobile manufacturing functions it is possible to combine the LAC curves for each function in order to examine the overall LAC for a car firm as a whole and reveal the MES for the industry. Thus, from the evidence presented in the sections above, it is suggested that the MEPS for each function is as follows:

- BIW LAC(B): MEPS = 500,000 units a year.
- Powertrain LAC(P): MEPS = 1 million units a year.
- Final assembly LAC(A): MEPS = 350,000 units a year.
- R&D LAC(R): MEPS = 600,000 units a year (production), 1,200 personnel, three platforms, five models.

To arrive at an MES output figure for the ideal firm it is necessary to calculate it from the lowest common denominator amongst the processes. This results in an output figure of 21 million units, a result that is nearly half the total global production for 2005. However, each process has a reasonably flat range alongside the MEPS output figure, so a visual analysis of the LAC cost curves can indicate an acceptable point for the MES. Figure 3.12 does this by combining the separate LACs for the complete, integrated firm. LAC(B) has been positioned high relative to the cost axis since it is the most capital intensive process, with LAC(P) slightly further down since it includes additional labor in assembly work. Final assembly, LAC(A), where all the parts arc brought together, has been positioned lower down the cost axis because it is the most labor intensive. R&D is the most dependent on human assets and so the LAC(R) curve is the flattest, revealing a shallow slope of increasing returns to scale as team effects and economies of scope bring benefits.

Visual inspection of Figure 3.12 seems to indicate that output at 600,000 units a year results in an acceptable and achievable low level of costs for

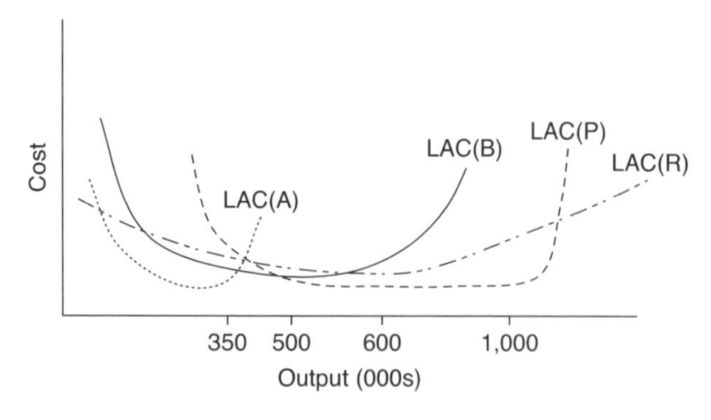

Figure 3.12 LAC curves for a prototypical integrated automobile firm

the combination of functions. It may not represent the absolute optimum, but with only the final assembly plant needed to be duplicated to match the output then it offers a pragmatic solution for the industry. This also accords with the results of the survivor analysis which indicated that the minimum output range for survival was 0.5 to 1 million units a year. This results in a company structure as shown in Figure 3.13.

In lieu of any manufacturer achieving 21 million units a year, then an output of 600,000 units a year represents, to a reasonable and pragmatic degree, the prototypical MES for the industry, known in this book as MES_P. A similar output for MES was also mentioned in an interview for this book with automotive analyst Mr Yamaguchi, who put the figure between 500,000 and 700,000 units a year.

Making use of the horizontal tendency in the LAC curves for each process it is possible to have plants working at their full designated capacity, albeit one that is not at absolute measures of the MEPS. In this way the schematic shown in Figure 3.13 would represent a prototypical model of an automobile firm with an R&D function at the optimum, a BIW function being slightly beyond the optimum, and the powertrain and final assembly functions being operated slightly suboptimally relative to absolute measures of the MEPS. Any cost penalties would be minimized by planning for these outputs, it being common, for example, to find final assembly plants operating at around 250,000 units a year. However, the greater cost penalties emerge when the installed production systems are themselves underused, resulting in a relatively rapid rise in costs on the prototypical firm's SATC curve ($SATC_P$). Figure 3.14 shows how a prototypical firm that planned for the appropriate level of output for achieving MES_P would be vulnerable to fluctuations in demand in the short term.

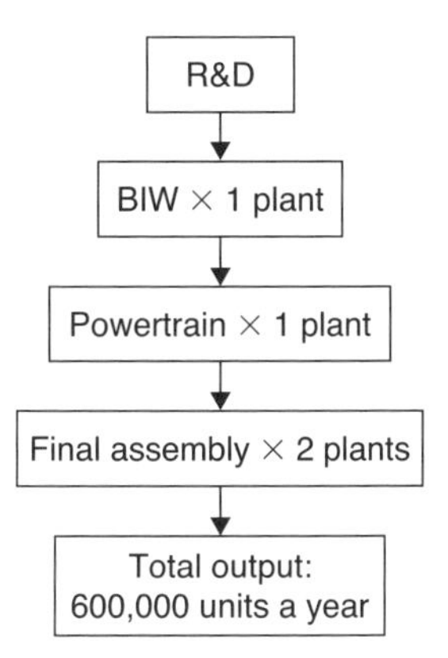

Figure 3.13 Automobile company exploiting acceptable economies of scale

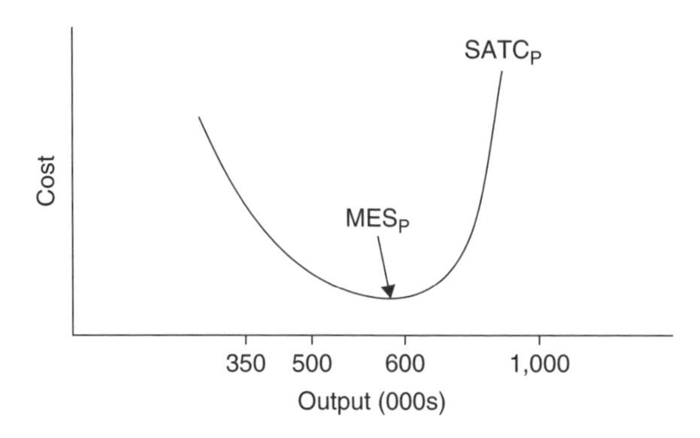

Figure 3.14 SATC curve for a prototypical integrated automobile firm

Figure 3.14 implies that firms would incur the greater cost penalties by operating above the minimum, but this can be managed by controlling demand in the form of an order bank or a customer waiting list. Suboptimal production presents more difficulties in the short term and demand would be managed using various marketing tactics such as price incentives, a cost that would then have to be balanced with the cost advantage of raising production back towards the MES_P output.

The sensitivity of an MES_P to output changes has been foreshadowed in the survivor analysis of the world automobile industry, which found that firms producing less than 1 million units a year were in a weak position, and indeed this could be extended to manufacturers producing less than 2.5 million units a year. At the same time it was difficult to reconcile this with the continuing survival of certain firms within those output ranges, Subaru being a notable example. This is clarified by the sensitivity of the MES_P to output changes which would motivate a firm to diversify risk by finding higher levels of output and wider model ranges. The survivor analysis found Toyota to be the industry leader – though one with an output that is nearly 13 times that of Subaru and with nine times the number of models. If a Subaru model, such as the Impreza, were to suffer a catastrophic failure in the market such that sales fell by 50 percent then the 60,000 unit shortfall would reduce total production of the company by 10 percent. If Toyota were to suffer the failure of a model with the same loss in production then the reduction in output would be barely 0.008 percent of the total. Furthermore, with a total of 81 model variants on offer it is likely that much of the demand could be transferred to a similar Toyota model rather than being lost to a rival manufacturer.

Should such a shortfall occur, Toyota could absorb it by reducing output at one plant while leaving the rest operating at full capacity. This is because Toyota does not manufacture from plants that are 13 times the size of Subaru's; instead it has a multitude of plants of a comparable size and up to 52 manufacturing companies overseas, even if not all are engaged in automobile manufacturing (Toyota 2007b). Survivor analysis has demonstrated the expansion of production at Toyota, as well as other industry leading manufacturers. As each new plant is brought on line the company has advanced in steps that can be measured in units of MEPS. In other words, a new assembly plant that is constructed in strict accordance with the MEPS will bring an increased capacity of 350,000 units. For a company like Subaru this would represent an increase in capacity greater than 50 percent but for Toyota it amounts to just a 0.05 percent increase in capacity.

Since a new plant must be served by other functions the firm will expand in units of the MES, and in the case of the MES_P this was found to be around 600,000 units. This may not be strictly necessary if the firm has access to different factor endowments such that it is able to accept different levels of output. For example, my research has found that Toyota engine plants had an output of 500,000 engines a year and final assembly plants had an output of 250,000 units a year. The firm could thus expand in steps of 500,000 production capacity units, although actual rises in output may not be this dramatic, as it takes time to bring a new plant up to full working capacity. Nevertheless, this clearly demonstrates that increases in capacity for larger firms are much less risky than for smaller firms.

If firms in the automobile industry are expanding in units of the MEPS to reduce risk then as they expand output the SATC curve will flatten out in recognition of the reduced variation in cost with fluctuations in output. However, I have shown that R&D costs rise significantly as firms gain in size. Figure 3.15 therefore shows the SATC curves for three types of automobile firms:

- $SATC_P$: prototypical firm, optimum output at an MES_P of 0.6 million units.
- $SATC_I$: intermediate firm, optimum output at 1.2 million units.
- $SATC_L$: industry leading firm, optimum output at 5 million units.

The LAC curve is scalloped around the three firm sizes, although it should be noted that other intermediate sizes of firm are available according to the multiples of the $SATC_P$. It should also be noted that firms smaller than the $SATC_P$ have not been shown but it is assumed that the LAC curve rises steeply for these lower output quantities, thus they suffer unsustainable cost disadvantages in production compared to firms at, or above, the MES_P.

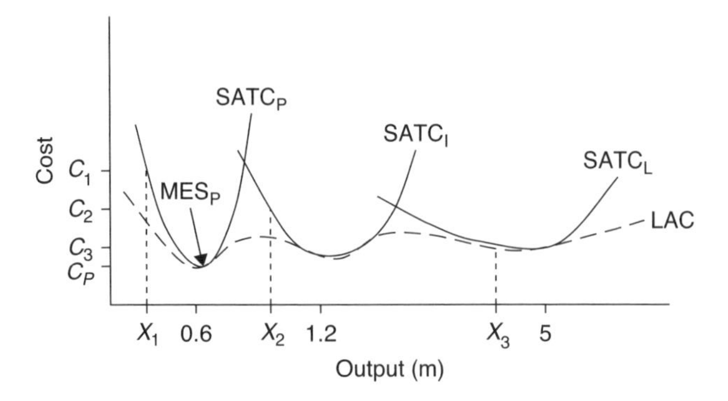

Figure 3.15 The SATC and LAC curves for automobile industry firms

The LAC curve shows rising costs after the MES_P on the SATC_P because R&D increasingly experiences diseconomies of scale. In this example, each firm is shown to suffer the same amount of production shortfall from its minimum cost position, the reduced output denoted by points X_1, X_2, and X_3. Since the SATC curves become increasingly flat with the growth of the firm, the costs disadvantages that relate to the shortfall become less marked. The costs associated with the MES_P are shown at C_P, but for output X_1 on the SATC_P the costs rise to C_1. For output X_2 on the SATC_I, the costs are shown by C_2, meaning that the production shortfall has had less of an effect than it did on SATC_P. The least affected is the production shortfall on SATC_L, shown at X_3, which incurs costs C_3. Not only are these costs barely distinguishable from what they might have been at the minimum cost output for this curve of 5 million units, but they are very close to the costs associated with the MES_P for the industry. This shows that although larger firms incur cost penalties in R&D and, moreover, are unable to achieve their minimum cost output because it would be almost impossible to have a multitude of factories all operating at full capacity at the same time, the shallow slope of the SATC curve means that they are able to remain close to the optimum for the industry despite variations in output.

It is this ability of larger firms to remain reliably close to the minimum costs for the industry that allows them to benefit from an influx of revenue. For example, although Subaru is a profitable firm, its net operating income for the 2007 financial year (ending March 2007) amounted to £44.4 million (US$88.8 million), whereas for Toyota, in the same period, net operating income amounted to £7.0 billion (US$14.0 billion) (source: company reports). Expressed as net operating income per unit, Toyota earned £909 (US$1820) per unit and Subaru earned £74 (US$148) per unit. Of course, this is a single sample and other years might show less of a difference;

nevertheless the difference is wide enough to be indicative. It should also be noted that there are cases where large firms, despite the diversification of risk, suffer a multiple failure which incurs heavy production and financial losses which can only be addressed by multiple plant closures to realign the company on a new SATC curve; the restructuring that has taken place at companies like Ford and GM would be a case in point.

3.3 Conclusion

In this chapter I have described how the presence of economies of scale for the four main functions in automobile manufacturing is expressed as the MEPS for plants and as the MES figures for integrated companies. Survivor analysis did not conclusively identify an overall MES for the industry within a growing market, although it did suggest a minimum output range for survival and reveal a trend by the largest manufacturers towards ever higher outputs and dominant market shares. Instead, established companies were investigated for the MEPS related to individual processes. Since each function uses a distinct production technology it would be unlikely that each MEPS output figure could easily be matched with that of the other functions to arrive at an accessible MES output for the firm as a whole. Instead, the flat ranges in each of the function LAC curves allow a rather more pragmatic MES output to emerge, that of the MES_P. This was found to occur at around 600,000 units a year. However, companies operating at such a level of output are sensitive to changes in output which lead to rapidly increasing costs. Larger companies are less susceptible to the onset of these increased costs since they can diversify the risk across larger numbers of products and plants. An output higher than the MES_P is therefore a strategic move to reduce risk rather than a search for further economies of scale. Indeed, rising R&D costs beyond the MES_P indicate that larger firms may be suffering diseconomies of scale but ones that are countered by the flattening of the SATC curve as output capacity increases.

There is nothing in the foregoing to suggest that a company is compelled to combine all four of the automobile manufacturing functions. In attempting to bring together the LAC curves for each function a company might consider instead focusing only on those functions where it felt it had the advantage. Generally, though, this does not happen and only the smallest of manufacturers operate with only three or less functions. Since this cannot be explained by economies of scale, which are concerned with technical efficiencies, in the next chapter of this book I will investigate the reasons behind vertical integration of the four functions.

CHAPTER 4

Structuring the firm

When a function is internalized the relevant economies of scale then become the responsibility of the parent company, but this does not have a material effect on the quantification of the cost benefits of scale. For the purposes of the generic model being constructed by this research, the argument is clarified by keeping the two concepts separate. The preceding two chapters have set down the principles of scale and demonstrated how these determine the size of the automobile industry. In particular that, for a given production process using a specified production technology, only a predefined production output range can make the most efficient use of the factors of production. In Chapter 3 I demonstrated these output ranges for each function based on data gathered from the literature, industry sources, and fieldwork interviews.

In order to demonstrate how companies in general are structured in this chapter I will investigate TCA (transaction cost analysis) as a mechanism for explaining the inclination of firms to integrate vertically. This will include the cost advantages of internalizing functions and how they might be managed within a governance structure. In the next chapter, Chapter 5, the automobile industry will be used as an example of how structural considerations can lead to a standard prototypical organizational form of vertical integration for a car firm. As a result, the economies of scale data in Chapter 3 can then be drawn on to provide the sizes of plant that are included within each internalized function. Since the considerations of size and structure prescribe a basic type of automobile company, one with defined limitations and potential, I will term this type an "automobile industry paradigm."

4.1 Company structure

I will now examine TCA theory as a structuring approach to the economics of business relations. This concerns the "make-or-buy" decision: whether to produce in-house as part of a firm's hierarchical structure, or else to buy from another firm through the mechanism of the market. The market relationship is governed by bilateral contracts formulated *ex ante* but which cannot account for all future contingencies. Consequently, the market relationship may suffer from *ex post* opportunism or hold-ups when benefits accrue disproportionately to one of the parties. The economic costs of this can be likened to a friction between the parties, one which might be termed "economic friction" in this situation. In order to overcome this friction one party may vertically integrate the other into its internal hierarchical structure. In a mature and global industry, where these business relationships may be a common feature, TCA implies that there is an optimal structure for firms that are using the same production technology.

A firm's structure is designed to minimize the costs of company functions working with each other. The purpose is for a firm to achieve the same, or better, minimum costs as its rivals. As David and Han (2004, 40) summarize it: "The theory's central claim is that transactions will be handled in such a way as to minimize the costs involved in carrying them out."

Cost is therefore the deciding factor behind "make-or-buy." In order to define when transaction costs are relevant David and Han (2004) list three ways in which transaction costs are manifested:

- Asset specificity: the degree to which assets may be redeployed.
- Uncertainty: the predictability of *ex ante* outcomes.
- Frequency: how often the relationship needs to be renegotiated.

This chapter will show that by a historical process of consolidation these considerations may culminate in vertically integrated firms based around their main functions. Since the degree of vertical integration has unlimited potential the term "fully vertically integrated" is a nebulous one. In order to confine the discussion to the core functions, the comprehensively vertically integrated firm will be said to be structured in accordance with the forthcoming proposal: the full-function model. This is to obviate comparisons between firms operating at different levels of scale which have the opportunity to integrate with a multitude of peripheral functions. Vertical integration here is not about absorbing quantities of external resources but about internalizing the core functions that define an industry.

Vertical integration and the full-function company

When a company takes more responsibility for production processes by internalizing the supply chain it is said to be increasingly vertically integrated. A vertically integrated manufacturer is one that has internalized all the major functions of production, although subcomponents and raw materials may still be supplied from external sources. Horizontal integration is related to economies of scale, such as the scope of products and the quantity of output, while vertical integration is concerned with the number of production stages included within the firm (Cabral 2000).

Vertical integration is a relative term because it depends on the degree to which processes have been internalized, and so complete vertical integration would be difficult to achieve since this would include even the processing of raw materials. The minutiae of vertical integration are not the focus here, though the exposure to the concomitant economies of scale is relevant. For this reason, companies that operate the main manufacturing functions, and are therefore exposed to the attendant economies of scale, will here be termed "full-function" rather than vertically integrated. In effect, the production process is being treated as interconnected modules of discrete production activity, each with its own inherent economies of scale, as described in the previous chapter.

Figure 4.1 illustrates the company boundary of a typical full-function company with four modules of production. The product originates in the design function and is put together in the assembly function. The product is made up of two components manufactured in two technically unrelated processes, Process 1 and Process 2, though a more complex product would require more processes. Each process is analogous to a production plant, although in practice the plants may be multiplied. The final products are released to the market which is, of course, external to the functional operations of the company, though information from it is fed back to the company's design function by way of a marketing capability.

A furniture factory, for example, might have a process for manufacturing wooden frames and another for the soft fabrics, there being no technical

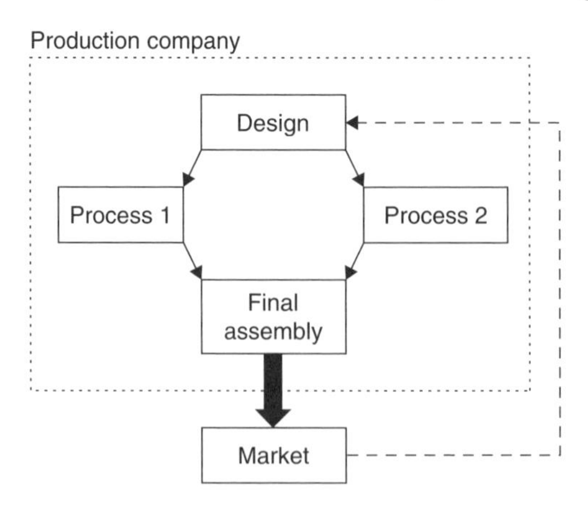

Figure 4.1 Typical full-function producer

commonality between the two. The manner in which the major functions relate to each other is a key element in the structuring of the company, initially as to whether the functions should be internalized and then as to the level of output for each. Concerning internalization, TCA is an instructive method for comprehending the resultant vertical integration structure of the company. Outside of the company's boundary lie the suppliers with which the company has a market-based relationship.

Component supply industry and functional relationships

A multistage production system involves inputs from a myriad of component suppliers. Even for a full-function firm operating in multiple core areas there will be component suppliers contributing at the subfunctionary level. These range from small individual parts to larger subassemblies that are built up from a collection of components. The completed subunit is then fed into the main assembly line.

The firm is in a vertical relationship with a supplier if its contribution is at a prior stage in the sequence of the production transformation process, thereby enabling the firm to complete its work. The part being supplied would therefore have to make a unique contribution to the production process, whereas if it were duplicating a current activity at the firm the relationship would be one of "tapered integration" (Cabral 2000). Between different suppliers the relationship is horizontal if they are producing broadly the same product. In terms of integration, it is vertical when a producer internalizes a supplier of complementary parts, and it is horizontal when the absorbed company is supplementing a current capability. Horizontal integration is particularly pertinent in international growth, when a foreign supplier is absorbed in order to secure a component supply at the new location in place of components sent from the home base.

The varying levels of integration between suppliers and producers are illuminated by the range of relevant factors, as stated by Odaka *et al.* (1988). This includes the technological distinctiveness of the part, production technology, demand, and economies of scale. Components may use a unique technology or may be an item that is common to other industries. Walker and Weber (1987) listed three costs that arise in the buyer–supplier relationship:

- Switching costs incurred by the buyer when changing supplier.
- Adjustment costs incurred by the supplier when changing the product or output.
- Transaction costs due to opportunistic behavior by the supplier.

For generic components the firm has purchasing power and can switch suppliers for a relatively low cost. This might be termed a low dependency relationship and one that can be controlled by the mechanisms of the open market. High adjustment costs occur when the supplier has specialized to the buyer, thereby introducing specificity in the transaction, but the supplier can also use this position to hold the buyer hostage to the relationship. For example, in fuel cell vehicle development Honda first purchased fuel cell stacks from Ballard of Canada for the FCX V1 in 1999, but began developing its own independent technology at the same time for the FCX V2 vehicle. Perhaps one of the most famous cases, though, was that between GM and Fisher Body during the 1920s when the car industry moved into the Budd Paradigm style of unitary body production. This will be discussed more fully later in this chapter.

Not only can opportunistic behavior and adjustment be seen as categories of transaction costs but so too can switching costs. TCA is concerned with all the costs that arise in managing the relationship between buyer and supplier, culminating in the make-or-buy decision. The transactions between buyer and supplier are governed by contracts, but even the most comprehensive of contracts cannot account for all eventualities. Transaction costs therefore arise due to the element of risk and uncertainty that create economic drag or friction, expressed as costs.

Transactions as mechanical friction

The transaction cost model involves the transfer of knowledge and resources under a system of ownership controls. Simply put, the firm will expand the scope of its activities to a point where the cost of an activity carried out within the organization equals the cost of conducting it with an external organization. The purpose is to minimize costs across separate technologies either by integrating activities or contracting with external organizations (Williamson 1998). These costs are not necessarily quantifiable: "Unlike production costs, transaction costs are very difficult to measure because they represent the potential consequences of alternative decisions" (Klein *et al.* 1990, 197).

Dealing with another organization brings with it additional costs, such as the information search to find the appropriate partner, bargaining costs in arranging the relationship, and administration costs in the course of the relationship (Viitanen 2002). The relationship is governed by mutually agreed contracts and the externalities can be summarized as follows:

- Search costs.
- Contracting costs.
- Coordination costs.

Although IT improvements have reduced relative costs of external resources, this is a contributory factor and will not necessarily eliminate altogether the costs of administering an external relationship.

TCA focuses on the links between different functional activities. The point of delivery represents the interface between the supplier's and the producer's manufacturing processes, regulated by formal statements of the bilateral relationship. When the supplier delivers the components right up to the assembly stations this brings the point of delivery into closer proximity but it does not change the fundamentals of the supplier–producer relationship. Functional integration would only occur if the assembler found, during an investigation into the costs of this relationship, that it could achieve lower costs by absorbing the component production within its own corporate structure. Levy (1985) was of the opinion that integration was a logical outcome of working in close proximity because it realized flow economies through a highly transaction-specific relationship. However, this does not necessarily change the physical nature of the component production and supply, since JIT supply is not reliant on internalization, although it may open possibilities to subsequent innovation. When TCA leads to increased integration it is done to reduce the costs of contracting with an external organization, not to change the nature of the processes.

Choosing to integrate the processes within the firm does not mean that there has occurred a market imperfection – the resources are still available externally – but it does represent a failure of the market to provide the most cost effective structure. In essence, the industry and the product remain unchanged. Williamson (1981) made the point that transaction costs are the economic analogy of mechanical friction between moving parts – they are the drag that occurs when transferring between activities – and this has been reiterated by subsequent researchers such as David and Han (2004). The activities are technologically separable goods and services, the transaction being the transfer process. The point could also be made that technological separation implies that different economies of scale are available for these activities. However, the interest here is not with the internal costs of each process but the costs arising between processes. Like David and Han (2004), Williamson (1987) put transaction costs into three dimensions:

- Frequency.
- Uncertainty.
- Specificity.

Williamson placed the emphasis on the last two. Contracts serve to control the transfer between the principal and agent, for example between the assembler and the component supplier. Uncertainties result from "human nature as we know it" and the risk is heightened by the degree of commitment, or specificity, of the assets (Williamson 1981, 553). While the market offers the freedom to find business relationships, contracts are needed to control the form of the relationship, since Williamson assumes that human nature will play a central role.

Contracts and the market

The behavioral assumptions Williamson (1981) had in mind are the bounded rationality of agents acting within their available knowledge, and the tendency towards opportunism when new knowledge emerges after the contract has been agreed. If all relevant information were known then complete bilateral contracts could be written *ex ante*, but this not being the case no contract can be formulated to cover all possible contingencies. This is exacerbated by the human desire to exploit subsequent advantages when they accrue asymmetrically to one of the parties, giving rise to an *ex post* opportunist situation where one of the partners can exert power not accounted for in the original contract. One form of opportunism is the "hold-up," where the owner of the relevant assets threatens to withhold supply in order to extract terms more favorable to its operations. For Williamson, opportunism is the main reason for attempting to regulate business through contracts or vertical integration. "But for opportunism, most forms of complex contracting and hierarchy vanish" (Williamson 1993, 97).

Williamson was perhaps displaying an overly negative opinion of human nature, and he conceded that transacting parties are usually "well-socialized" enough to economize to a degree on transaction costs. Furthermore, an incomplete contract is also one that has an inbuilt flexibility that might yield unforeseen innovations. A rigid contract can bind two firms together when economic efficiency would be better served if they dissolved their relationship (Tirole 1989). This may even be unwittingly entered into, as Williamson (1993, 100) suggested: "opportunism is more often suppressed unknowingly or selectively and that, once done, the ramifications are rarely assessed."

For Klein (2000) there is another, implicit, way in which the incomplete contract can be confined. If the benefits of reneging on the contract are offset by damage done to the offender's good name, what can be termed its "reputational capital," then the contract is costlessly self-enforcing. This is an attractive proposition for a company because internalization also has costs. When a company absorbs a supplier it is, by definition of the relationship, expanding into an industry in which it previously had little or no experience and may push the firm into management diseconomies. Although the information search costs in seeking suppliers are being reduced by such a move there then appears an opportunity cost in forgoing the engagement of suppliers in the future that might have been more suitable. In a sense, the producer becomes locked-in to a supplier that may not provide the greatest benefits. This may be accentuated during a time of fast moving innovation, the period of uncertainty described by Williamson, when an alternative supplier may unveil a development of considerable impact.

The role of the market

The foundation of the dichotomous decision making, integration of resources, or finding those resources in the open market is the nature of the market itself. However, there are difficulties in defining precisely what shape the market takes. Ankarloo and Palermo (2004) restate Williamson's assumption on the primordial existence of the market, with zero transaction costs acting as a reference point. Since such a utopian state can only be aspired to, due to the human frailties described by Williamson, then transaction costs mean that relationships have to be regulated by contracts. The concept of the firm is then defined by its boundary of contracts linking it to the market. This is suggestive of a sense of evolution from the pure market to one that deals with inevitable, if unpredictable, market failures. Ankarloo and Palermo dispute the dichotomy, pointing out that the market basis does not explain the existence of economic institutions like firms, since Pareto efficiency can be achieved by a centralized and formal allocative model that a government-controlled economy would provide. This is a problem for Williamson's theory since he bases his theory of the firm, and transaction costs, on the failures of the market. If an alternative to the market exists then there are reasons for firms to exist other than to deal with transaction cost problems and market failures.

Ankarloo and Palermo's Marxist polemic does not actually rule out a role for transaction costs but they do introduce a third element to refute the market–firm dichotomy. This is the regulatory function of the state to legislate a framework for markets and the manner in which companies should be permitted to operate within them – yet the authors seem unaware that this is simply an institutionalized version of the contract. Moreover, since the functions of the state need to be paid for, any state promulgated regulations bring with them their own costs, thus reinforcing the idea of market failure and transaction costs. Klein (2000, 127) illustrates this when he states that referral of a contract in dispute to a third party introduces a time lag and that the need to communicate the minutiae of the contract results in "increased noise." These costs are avoidable when a self-enforcement mechanism exists, such as when the renegade firm risks injury to its reputation or "transactor's reputational capital." Klein notes that this is particularly effective with large contracts.

To take the contracting process out of the hands of the contracting parties and utilize the services of a regulatory intermediary is simply to politicize the existing process – it does not create a third alternative to the market and the firm. It is a fundamental point that for a government to regulate a system of contracts, which is in itself already a form of regulation, is either to add another layer of regulation or at least to create the forum where the contracts are formulated – but this does not provide an alternative to the necessity of making contracts. It hardly matters whether the primordial market ever existed or is an "expositional convenience" used by Williamson (1975, 20). Conversely, it would be quite impossible for the concept of the firm to predate the market unless all economic activity had begun with one universally integrated firm. Williamson states that it would even be possible to start from a theoretical position of primordial central planning, the point being made is that economic friction caused by bounded rationality manifests itself as transaction costs. The purpose of the firm, then, is to internalize the transactions and so reduce the costs, thereon expanding until the costs of the internal transactions are equal to or exceed the costs of external transactions between firms. The market is simply the external relationship between firms, not an institution in itself. Moreover, the market is not defined by the medium of the transaction, although money tends to reduce transaction costs in comparison to barter since it is usually mutually acceptable (Williamson and Wright 1994).

The imperfect contract

Williamson (1975) looked at the market as a theoretical starting point, but he could not define its exact nature because, as the antithesis of a contractually controlled system, the market is unstructured and unpredictable. If the market were predictable then contracts could be perfectly formulated to cover all contingencies. The choice companies face is between internalization of resources into an integrated hierarchical firm or finding those resources in the market and regulating them by contracts. The transaction-cost approach is not derived from some poorly defined notion of the market; though it does have its varied antecedents, namely organization theory and contract law, in addition to economics (Williamson 1981, 550).

These three are indivisibly interrelated, a contract being written by participating organizations in order to gain some control over the economic friction between them.

Concerning contract law, Klein (1980, 356) underlined the basic problem of contracts, that "complete, fully contingent, costlessly enforceable contracts are not possible." This is stated without direct reference to the market because contracts are made as security, however limited, against the other party's failure to act in accordance with expectations. Slater and Spencer (2000) discuss the difference between uncertainty, which is unknowable, and risk, which may be calculated for by the transacting parties. Yet the insurance industry is able to calculate premiums without having to make a fine distinction between the two concepts, so it could be suggested that from the perspective of transaction-costs theory the difference is mainly semantic. In any case, Klein (1980, 357) did not express a need to distinguish between the two: "Contracts can be usefully thought to refer to anticipated rather than stated performance."

Two reasons are given for no contract being complete: firstly, that not all contingencies can be anticipated; secondly, that performance is difficult to predict. Specific reference is even made to Williamson's warning about "opportunistic behaviour." This is a reminder that transaction costs and contracts are formulated with reference to the behavioral issues of economics and organizations, not the conditions of a theoretical original market. Klein (1988, 202) took this further, showing that long-term contracts, while they are partially enforced by reputational capital, with a large enough shift in the demand or supply characteristics will move the "contractual arrangement outside of the self-enforcing range." Although investments specific to the business relationship require long-term contracts for security, far from enforcing the relationship this actually creates the conditions for a potential hold-up, albeit one that, *ex ante*, is of unknowable source. It is the nature of this specificity that is relevant to the necessity for vertical integration.

Specificity

A crucial feature of a contractual relationship is the degree to which any assets are specific to it, and therefore less easily employed elsewhere. Klein (1980) particularly focused on the risks involved when firm-specific investments are made, since the firm making them can be held hostage by its contractual partner, which can then exploit its advantage by appropriating the quasi-rent stream. The investing firm has more to lose due to the difficulty in transferring the investment, whereas the renegade firm has most to gain, at least in the short-term. Williamson (1981) also stressed this point, preferring the term "asset specificity," but supporting the view that increased specificity brings higher risks and therefore more significant transaction costs.

According to Williamson (1981, 555) specificity can occur in three ways:

- Site specificity.
- Physical asset specificity.
- Human asset specificity.

The specificity is analogous to commitment, since the value of the investment being made is higher for the relationship in question over any other. For example, the fast-food chain McDonald's operates restaurants with franchise arrangements. The main physical assets are the restaurant buildings with brand-specific architecture and interior design. Since these are important in projecting the appropriate corporate image the company retains control of them by owning them. This power is partially countered by the franchisee's power to withhold the service standards set by McDonald's, the franchisee having the potential to profit from cost savings that accrue to their part of the business but which are damaging to the corporate business as a whole (Milgrom and Roberts 1992). McDonald's is therefore careful to put in place a system of incentives and penalties to support the contractual relationship. Lieberman (1991) found strong empirical support for integration by firms that carried significant sunk costs due to asset specificity and anxiety about being "locked-in" to a business relationship. Lyons (1995) also found asset specificity to be predictive of vertical integration in UK engineering firms.

Additional specificities that are derived from the three mentioned above have also been put forward. Williamson (1983) included dedicated assets which are generic in nature but specific to the particular transaction and which might have been put in place in expectation of new business. This being the case then, as far as the transaction in question is concerned, they have the same risks for the supplier as physical asset specificity; should the deal fall through the assets have less value when applied elsewhere. Joskow (2003) puts forward the idea of intangible asset specificity as would be contained within a brand image, giving the example of McDonald's. Again this concept is derivative and aspects of it appear in human asset specificity (service) and physical asset specificity (restaurant decor). It is also covered by the reputational capital described by Klein (2000).

Masten *et al.* (1991) added temporal or time specificity, where a supplier of a time sensitive product gains progressively greater power over the buyer as the deadline looms. It is notable that defense contracts in the UK often detail penalties to counter this hazard, the most recent example being the overruns on the Nimrod reconnaissance aircraft, which gave rise to a £500 million (US$920 million) penalty against the manufacturer, BAE Systems, by the British government (*Sunday Times* 2006). High temporal specificity also occurs on a sequential production line where a stoppage at one place causes the entire line to halt. Joskow (2003) believes this to be a subcategory of site specificity. Indeed, specificity as a concept is most comprehensively covered by the site, physical, and human specificities originally set out by Williamson (1981), the derivations being useful in explaining the nuances of particular instances.

Site asset specificity

Site specificity anchors an industrial activity in a particular location due to the pre-existing characteristics of the site. This can be taken to mean the natural resources as a factor of production or the industrial environment. As Joskow (1985, 38) described it: "buyer and seller are in a 'cheek-by-jowl' relation with one another."

Historically, industries were usually tied to a particular location due to the reliance on natural resources and poorly developed transportation. Sheffield rose as a steel town due to the proximity of iron, coal, and local rivers for water power. Once the industry was established it took on Joskow's more sophisticated definition of site specificity based on external economies of scale, attracting developments in steel making due to the established industrial infrastructure as much as the natural resources, "reflecting *ex ante* decisions to minimize inventory and transportation costs" (Joskow 2003, 14).

Nuclear power stations also need a readily available supply of water but they tend to be located in less populated areas. This avoidance of a particular factor, rather than attraction to it, might be termed "negative site specificity." Once operating, though, a second nuclear power station can be built in juxtaposition, yet the two power generators are not in a buyer–seller transactional relationship. The proximity of the second station is due more to reasons of economies of scale and political issues, including the original negative site specificity.

With advancements in transportation links, political pressures can often be the deciding factor in site specificity. UK government policies have influenced the choice of location, from intervention by way of the Industrial Development Certificate of the 1960s (Adeney 1988) to inducement by way of financial incentives later in the 1980s and beyond (Garrahan and Stewart 1991). In effect, this altered the price of factor inputs. However, the 1960s showed that while government policy could be instrumental in locating manufacturers it was not the sole motivator, and the long-term success was not one that could be reliably predicted. Jaguar avoided having to move by taking over Daimler's Browns Lane plant in 1952 (Underwood 1989). This plant continued in production for a further 53 years before it was found to be uneconomic in comparison to the other Jaguar sites at Halewood and Castle Bromwich, at which point it was closed as a production facility.

Physical asset specificity

Physical asset specificity concerns the degree to which capital investments are firm specific. Williamson (1981) told us that asset specificity is the most important element in transaction costs: the less specialized the assets, the fewer the hazards. When specialized facilities are installed then the relationship between the buyer and seller becomes "locked-in" with both parties interested in the longevity of the contract. In a survey of hospitals, Coles and Hesterly (1998) concluded that specialized equipment that could not be transferred to other hospitals, because it was configured for a service unique to one hospital, was defined as a specific asset. Physical asset specificity can therefore be contingent on the service being offered, it being possible to offer a reduced standard of service with standardized capital equipment. However, Williamson's perspective on this is that once the more specialized medical equipment is installed it is then in the interests of the surgeons to remain with the equipment, based on the rational assumption that this would result in a higher quality of service. Coase was not so impressed by the impact of asset specificity, noting that businessmen he spoke to in 1932 did not attach great importance to it since they could arrange contracts that took care of their concerns. He claims that "contractual arrangements were able to handle the asset specificity problem in a satisfactory manner" (Coase 2000, 18).

The views of Coase and Williamson are not necessarily irreconcilable. Williamson is taking a holistic perspective, *ex ante* to *ex post*. Entering into an asset specific transaction carries higher risks than if standardized, transferable assets were used. Early on in the relationship hold-ups can be enacted as the holder of the assets is able to exploit their position to extract concessions from the purchaser. Once the relationship is stabilized, Williamson believes in the longevity of the relationship, a phenomenon that Coase observed *ex post*. Specific physical assets appear to create high entrance and exit costs which have to be assessed *ex ante*, but this also brings stability to the relationship once all costs have been accounted for, and as Williamson stated (1993, 105): "the wise prince is one who seeks to both give and receive credible commitments." This also acts as a reminder that transactions and contracts are aspects of human relationships, indicating that human asset specificity might have a crucial role.

Human asset specificity

The unique knowledge and skill that is dedicated to the transaction is of human origin, consequently known as human specificity. This has received less attention than physical asset specificity, probably because it is less quantifiable. There are, of course, occasions during industrial disputes when the withdrawal of labor can bring production to a halt. However, it is only by the nature of strike action that replacement labor is not swiftly employed. In many other situations, when labor becomes unavailable it is possible to find other labor resources within a reasonably short period of time. It is because of this that labor is seen as one of the variable costs of production, rather than a fixed cost. Physical assets are, though, viewed as fixed and therefore physical asset specificity is seen as instrumental in longer term operations. The problem with this view is that it neglects to account for the root cause of the vulnerability that results from physical asset specificity. This is the opportunism that is practiced by the human actors in the contractual relationship, which then manifests itself through the appropriation of the physical assets. It is by the nature of human frailty that the human assets are the foundation of opportunism, even if they exercise their power through the physical assets.

Although physical assets, even at their most generic, tend to have readily definable tasks, human assets can play a complex number of roles. Coles and Hesterly (1998) drew a distinction between human asset specificity, which links the person to a unique service, and general human capital, which is transferable between organizations. The example they provide is of assistants who provide a service tailored to the particular needs of the surgeons at a hospital, a unique service that might not suit surgeons at another location. Human asset specificity is therefore inherent to the supplier, and the buyer can exploit this if there is a credible alternative source of supply. The supplier would then find that the knowledge accumulated could not be transferred to another buyer and still obtain the same value.

Since the asset itself resides within the worker they are also in a position to cause a hold-up, often witnessed collectively in strike action.

Joskow (2003) gives the example of the special skills that aircraft designers accrue through experience, but it is quite feasible that these designers could withhold their labor if they believed they were not being adequately recompensed (Williamson 1981). This represents an opportunity for the "appropriation of quasi-rents" (Klein 1988, 199). Klein believes that this situation is peculiar to human assets, indeed that the entire make-or-buy decision is actually a choice between the physical making of products and the buying of knowledge, not between internal integration and outsourcing. A firm does not need to own its physical assets to use them for production, and Klein gives the example of a firm not owning the offices in which it works. Furthermore, the entire production transformation process is replete with physical assets, any one of which has some potential to be the source of a hold-up. However, whatever the extent of ownership of the physical assets, the human assets that operate them can never be owned in the same way. It is solely through owning the labor contracts that a firm can have a claim on the human assets, and the contracts detail the extent of those claims. This definition of output by contract is also applicable to physical assets, which is the basis of the outsourcing agreement, but the owner of the physical assets has additional discretionary powers over the asset.

Klein (1988) asserted that in a free world the human assets are at liberty to seek employment elsewhere, since humans retain discretionary powers over themselves, notwithstanding the details of the labor contract. This might threaten the sustainability of the firm, but as long as key personnel remain then the knowledge embodied in the human assets will perpetuate within the organization: "The employees come and go but the organization maintains the memory of past trials and the knowledge of how to best do something" (Klein 1988, 208).

Klein describes this as ownership of the firm's organizational capital, which is embedded in the human assets. The human asset specificity lies with the organizational capital and so vertical integration internalizes this organizational capital, even if the number of labor contracts compared to the market remains constant. The great advance for the firm is that it is no longer bound to *ex post* consequences of the contracts because any advantages that accrue disproportionately to the human assets also accrue to the firm through its ownership of the organization's capital. Klein (1988) believed that even if some employees can threaten a hold-up by resigning, the organizational capital remains in place with the key personnel that stay.

Like all TCA this is a make-or-buy dichotomy: internal flexibility still foregoes the flexibility of accessing more widely spread human assets in the market, so the choice is based on relative costs. Nevertheless, within existing labor regulations the integrated company can make adjustments to plant locations, production output, and other tangible assets because it owns the labor contracts of the workers that operate those assets. The transaction costs that the integrated firm saves are in negotiating and renegotiating these contracts at the same time as avoiding one overarching, yet rigid, contract between two separate firms. This concords with the views of Hart and Moore (1990) concerning the property rights theory of the firm, which is more specifically based upon the ownership of physical assets, where the authors conclude that an integrated firm can shed workers selectively, but when contracting with another firm it is the whole firm that must be "fired." However, by taking a strictly physical asset view of the firm, it is difficult for the property rights theory to account for the opportunism of human frailty: "Surely integration does not give a boss direct control over workers' human capital, in the absence of slavery?" (Hart and Moore 1990, 1150).

It may be the case that a firm cannot claim ownership of its human assets, but the degree of control is laid out within the terms of the contract. It is also unnecessary to have complete control over each individual employee as well as over the collective organizational capital. Klein's insight into the ownership of the organizational capital through the employment contracts deals with this control issue and has crucial importance for knowledge intensive processes such as design work. Where the knowledge is at the core of the firm's operations then it needs to secure the services of those human assets that carry the knowledge, at least in high enough numbers that the knowledge is perpetuated within the organization. If the value of this knowledge is great enough it can push other cost considerations, such as economies of scale, into a distant second place. Klein argues that the same principle can be applied to the physical assets since knowledge is required to operate these too. This may be particularly relevant when the production technology is in flux during a time of rapid development. Conversely, where the technology is mature and capital intensive, it is the switching costs of the physical assets that come to the fore, the human assets being more easily replaced. The implication of this is that companies made up of processes that are, separately, knowledge intensive and capital intensive will vertically integrate those processes for different reasons: the human assets to secure knowledge, the physical assets to secure access to production.

As Klein points out, the securing of human assets cannot be done by direct contract and so additional incentives become the key for securing access to them. Holmstrom and Milgrom (1994) explored the complexity of incentive systems within firms. Since the monitoring of output from human assets is "complex and costly" recourse is taken to monitoring the output from physical assets and basing incentive schemes on that. In this way workers can be rewarded for production output and so it is in the interests of the firm owner to secure a better bargaining position through being the owner of the physical assets. This becomes more problematic when the output is dependent solely upon the human assets, such as in design work. As Poppo and Zenger (1998, 859) showed, the difficulties in achieving this through a market mechanism encourage firms to internalize knowledge-based activities: "Markets, by contrast, lack the capacity for such managerial intervention: when measurement is highly problematic, markets simply fail."

Internalization is, therefore, a system for gaining control by the management through ownership of labor contracts. This internalizes knowledge generation into the firm's hierarchical structure and so gaining some measure of control over it. Any unforeseen benefits that emerge, such as new product innovations, accrue to the firm through its ownership of the labor contracts. Although this has to be balanced with the loss of access to innovations occurring externally, and by definition innovations are unpredictable and of unknowable source, internalization of human assets means that they can be treated more like physical assets. Output, though, is more difficult to quantify for human assets than physical assets and so economies of scale for human assets are analyzed differently. This, in turn, suggests that the manner in which company functions are integrated depends on the preponderance of either human or physical assets within that function. In the next section I will investigate how the functions can be brought together in a prescribed governance format.

4.2 Unitary and multidivisional governance structures

TCA highlights the make-or-buy dichotomy in terms of the decision of whether to internalize an activity into a vertically integrated functional structure or to transact for it in the market. Chandler (1990) viewed this as a strategy since it is an allocation of resources necessary for carrying out long-term goals. Once the strategy has been determined and the functions internalized, then a corporate governance structure needs to be imposed on those functions that have been brought within the company boundary. At the simplest level, Chandler (1977) observed that before the arrival of the railway in the US those firms that were involved in the transportation of goods tended to leave regulation of the business environment to government legislation. However, railway transportation necessitated careful coordination of traffic and this was heightened by the technological advances in the speed of the locomotives. It was therefore necessary to implement a form of management that was controlled from the center.

The strategic purpose of the first railways engendered a hierarchical and simple form of structure, the unitary or U-Form structure. Generally, the simplest strategy is likely to be one pursued by firms that are small, or singular in their purpose, and so provide fertile ground for this kind of structure. This may begin with the most senior manager, often the owner, taking responsibility for all the major decisions which are then passed directly to those who will implement them. The centralized format of the U-Form provides a hierarchical structure for administering the functions of the organization (Ferguson and Ferguson 2000). It is designed to reduce opportunistic behavior by individual employees through the establishment of collective bargaining that consigns workers to recognized groups, thereby removing the possibility of acting to their own advantage. Personal advancement is then offered in the form of promotion, with new entry being restricted to the most basic levels of job definition (Moschandreas 1994).

In the U-Form structure the functions have functional managers who gather information which they present to the CEO and await further instructions. The CEO therefore directly overseas each function and is responsible for the daily operations of the company as well as its strategic direction. This means that managers specialize according to their function and communication is conducted functionally, meaning that information passes up the hierarchy to the CEO but not across to other functions, except through the CEO. The system assumes that the CEO has enough experience of the functions to be able to make informed operational decisions, but it also implies that the functions will work together when directed to do so by the CEO. Moschandreas (1994) argued that in practice the U-Form is vulnerable to "empire building," where functional managers attempt to cultivate as much influence with the CEO as possible in order to further the interests of their departments or even their own personal interests at the expense of profit maximization.

The U-Form is considered applicable as long as the numbers of functions under the CEO are not too diverse and each function can discharge its responsibilities. However, should a company internalize a greater variety of functions then it becomes progressively more difficult to coordinate them as if they were facets of a single business. For example, a company might have two manufacturing functions, one producing pharmaceuticals and the other cosmetics. The marketing requirements for each might be so distinct that a single marketing function would not be suitable and a way would have to be found to divide its operations between the two manufacturing functions (Ricketts 1987). In this way diversification brings with it complexity and the need to coordinate the different functions, meaning that the CEO has less time for strategic decisions. One solution is to add layers of bureaucracy, but this can reduce the flexibility of the enterprise and encourage opportunism (Moschandreas 1994). Hill (1985) put forward the concept of the H-Form, a holding company with centralized profit collection but decentralized strategic, financial, and operational control. However, this does not seem to offer a governance structure as such and might be better conceptualized as an investment vehicle.

Chandler (1977) observed that diversified, multiunit enterprises developed a middle management to monitor functional performance and coordinate material flows, while the senior management evaluated information flows passed up from the middle management and so allocated strategic resources accordingly. This supplanted the role of the market between independent organizations such that "the visible hand of management carried on the functions hitherto performed by market mechanisms in American industry" (Chandler 1977, 377).

This form of structure was the multidivisional M-Form, a decomposition of the U-Form. At its most basic it comprises a general office to monitor the operating divisions and implement strategic allocation of resources. The M-Form is a multidivisional organizational structure of bureaucracy that separates strategic and tactical planning, delegating operational control to internal divisions, while a central headquarters retains strategic and overall financial control. This prevents divisional managers from diverting funds opportunistically to their own operations and frees senior managers from making detailed operational decisions for which they lack information (Freeland 1996a, 1996b). As Klein (1988) and Hoskisson *et al.* (1993) also noted, ownership of labor contracts means that any errant staff can be more easily dealt with than through the mechanism of the market.

Having intimate knowledge of the financial affairs of each division the headquarters can allocate and monitor capital more closely than could shareholders through the external capital market. This means that it is more efficient at allocating resources, although if economies of scope existed between the functions when they were in the U-Form structure then they will be lost in the M-Form. Furthermore, when administrative functions such as human resources and accounts are distributed amongst the divisions they may lose access to the economies of scale in their functions that existed when they were unified in the U-Form. The M-Form therefore has the advantages of being able to allocate resources, inhibit opportunism, and encourage innovation within functions, but it may also increase the costs of operating those functions.

Governance structures in application

The M-Form originally came to prominence as a system of bureaucracy used by Alfred P. Sloan to control company divisions. Sloan had made his reputation with the Hyatt Roller Bearing company, purchasing it in 1899 for $5,000 (£1,030) and selling it to William Durant in 1916 for $13.5 million (£2.8 million). The absorption into General Motors (GM) put Sloan in control of the disparate United Motors component supplier subsidiary, which he then reorganized into semi-autonomous business units. With the near financial collapse of GM in November, 1920, its president, Pierre du Pont, invited Sloan to set about a similar strategy of reorganization for GM as a whole (Chandler 1990).

Sloan identified the main problem at GM as being lack of control, a classic transaction cost issue, with high levels of inventory and a confused product line at a time of economic recession (Sloan 1986, 42). This was tackled by drawing together the separate divisions under the command of a centralized administrative structure to ensure more effective coordination and control. Conversely, rival automotive company Ford had taken a highly integrative approach that was supported by a rigid system of controls throughout the company which precluded a unified central control. It also exhibited an overdependence on existing structures to the degree that it remained inflexible in the face of GM's rise (Kuhn 1986). Henry Ford had originally shown great innovation in creating his company, but in fashioning it in the image of production flows he had failed to apply a strong corporate governance structure.

The form of loose vertical integration at GM was based on three principles (Sloan 1986, 50):

1. A degree of independence for each division so that it had a sense of its own responsibility and contribution.
2. Appropriate measure of return on investment for each division to reveal the level of its efficiency.
3. Centralized decision making on additional investments with regard to their utility to the group as a whole.

The second two items are concerned with financial controls, while the first recognizes the advantages of the human element, described by Williamson as being at the root of opportunism, but here it is presented as having the potential for a positive contribution. In other words, opportunism is seen as a positive attribute to be nurtured and so illustrates that the human element is a symmetrical risk with both negative and positive outcomes possible. Williamson (1975) put forward three strategic control systems inherent to the M-Form:

1. Internal incentives and control to align personal interests with those of the firm.

2. An internal monitoring system which reports to senior management and that is more comprehensive and prompt than an external audit implemented for the benefit of shareholders.
3. Resource allocation based on objective evaluation of profitability.

In practice, the division of responsibility in an M-Form corporation may not be as distinct as the theory sets out. Assuming that the personnel who are stationed within the headquarters have been promoted up the corporate hierarchy, rather than being trained purely as strategists, then it might be expected that they would wish to exploit their experience of operational matters. As Moschandreas (1994) pointed out, senior managers may become involved in operational matters rather than restricting themselves purely to strategy. Furthermore, they may not be interested in profit maximization for the benefit of shareholders but simply a satisfactory return. There is also the possibility of opportunism within an internal incentive system amongst divisional managers who are eager to protect their own interests. Finally, the divisions themselves may become so big that they take on the same disadvantages as large U-Form firms, necessitating further divisionalization.

In order to test the empirical evidence for the M-Form, researchers have looked at how it may be applied in new and different situations. Chang and Choi (1988) took a transaction cost approach to analyze company structure with regard to the state of the market. This research focused on Korean companies operating in the poorly developed domestic market, one that can be assumed to contain more market imperfections than would those in more developed countries at that time. Korean business groups, known as *chaebols*, exhibit vertical integration and diversification, these enabling them to pursue both economies of scale and economies of scope. Although business groups are not confined to Korea they have been remarkably conspicuous in that country, taking 40.7 percent of total manufacturing shipments by value six years before the study took place. Donaldson (2001) argues that the M-Form reduces uncertainty by internalizing markets, and this can be observed in Korean business groups. However, this uncertainty is based upon the poor technical development of the resource market rather than uncertainty created by high rates of innovation. Korean firms can be seen to diversify in order to create a market rather than internalize one that already existed. Added to this is the fact, conceded by the authors, that the Korean groups at the time of the study enjoyed substantial external government support.

Hill (1988) suggested two types of diversification: related diversification and unrelated diversification. If the diversification is related, i.e. the divisions work with each other in the same industry, then the organization is structured to differentiate the functions so that they may specialize, while also internalizing them to reduce transaction costs. Such a multidivisional form would need more operational control of the functions from the center, a governance structure Hill terms the CM-Form. Alternatively, if the diversification is unrelated then the purpose of the M-Form is to exploit the economic benefits of an internalized capital market. According to Teece control of capital in unrelated diversification is effective because the senior management is privy to financial information to an extent that external shareholders are not, and is able to monitor and act on investments more swiftly and cheaply: "Furthermore the M-Form creates its own internal miniature capital market which replicates external capital market functions and economises on transactions cost" (Teece 1981, 175).

The U-Form integrated firm also internalizes the functions of the market, although it does not attempt to replicate the capital market to the same degree. The M-Form, by separating the divisions, creates an environment where the divisions must appeal for funds based on their investment potential. It might be argued that the same processes could be achieved within a U-Form structure, perhaps with the CEO personally gathering and evaluating the data of different functions rather than expecting pro forma information to be passed up the command chain, but this is not a capability that is inherent to the formal definition of the U-Form structure. However, it does suggest that the distinction between the M-Form and U-Form is characterized by the formal reporting of data, rather than the separation of operational and strategic decision making.

Taking the operational perspective, other literary sources considered how production was operated within the two basic types of governance structure. Burton and Obel (1980) found that the M-Form was suitable for parallel processes, when they share the same resources, but that the U-Form was suited for sequential processes that were working on different stages of one production system. Furthermore, relatively small companies were found to be better served by the U-Form alone, whereas larger companies have a choice between the M-Form and the U-Form. For these larger firms the evidence suggested that the M-Form was preferable under all circumstances but most especially when diversified and undergoing high rates of growth. It is at times such as these that clarity needs to be brought to the reporting of data, but in a fast moving environment there is also the difficulty in separating operational and strategic decision-making processes.

Donaldson (2001) draws a similar comparison between the M-Form and the U-Form structures. The research was conducted within contingency theory which seeks to align the appropriate organizational structure with the three factors of environment, company size, and strategy. For example, the functional structure of the U-Form has cost advantages in simplicity and specialization, which then places a high priority on the pursuit of economies of scale. Alternatively, the M-form decentralizes control of the functions and allows them to operate as semi-autonomous business units, a strategy which is particularly suited to highly diversified product ranges.

Focusing on the three factors of contingency theory, Donaldson (2001) defines environment as being the stability of the firm's situation, a vertically integrated hierarchical U-Form structure having the advantage when there is little technological change, while a more flexible "organic" M-Form structure is better suited when the industry is in flux. This is in line with the findings of Burton and Obel (1980) that the M-Form is well suited to accommodating change. Considering the factor of size, Donaldson (2001) argues that employment of larger numbers of staff mean that tasks are more repetitive and therefore amenable to the setting of policies by a distant M-Form headquarters, while smaller staff numbers tend to be engaged in less routine tasks and senior management can be more personally involved using a U-Form approach. It might also be argued that routine tasks need less operational control, suggesting that some M-Form headquarters might have actually been U-Form but were released from the need for regular operational involvement, thus, by default, leaving the management to focus on strategic issues.

Donaldson's last factor, strategy, concerns the range of products being made available, a diversified M-Form organizational structure being better suited to a diversified product range. Donaldson claims that contingency theory permits a wider range of corporate structures depending on the level of certainty; and, indeed, if Chandler is correct that structure follows strategy, then there should be as many structures as there are strategies. There may, though, be limits to the degree of change the M-Form can contain, and some theorists have suggested that the business landscape has experienced a quantum change due to technological developments and the globalization of markets, leading to a new form of organizational structure beyond the M-Form. Doz and Prahalad (1991) argued that the M-Form of structure was inadequate for accommodating the wide disparity between countries, functions, and approaches to business when the company is a diversified multinational corporation (DMNC). Although the M-Form relieves senior managers of making quotidian operational decisions they are still reliant on a flow of information in order to make judgements about resource allocation. Although the information should be of an objective type it still requires a degree of interpretation, and this can be difficult when the data is from an unfamiliar source, perhaps due to geographic or technical differences, possibly undermining the validity of the resource allocation decisions.

Bartlett and Ghoshal (1993) redefined the M-Form as a systematic devolution of assets from responsibility at corporate and divisional levels, suggesting that a new bottom-up approach would allow greater entrepreneurial and strategic control of the divisions. This recognizes that an organization is fundamentally a social structure and that a focus on functional structures obscures the roles of the human actors. As a consequence, Bartlett and Ghoshal argued that the three levels of management (top, middle and front-line) should no longer amend information from strategic to operational as it descends the corporate hierarchy. Instead, the front-line should become the source of entrepreneurial activity, the middle should coordinate the divisions, and the top management should create the overall purpose while in return challenging the status quo. This has particular ramifications for globally spread companies where the lines of communication may be too long for effective strategic control from the center to the dispersed divisions. As Freeland (1996a) pointed out, the sanctions and incentives by which the M-Form headquarters controls the divisions is dependent on the quality of the information passing between them. Bartlett and Ghoshal's "managerial theory of the firm" allows local divisions of DMNCs to exploit local resources and opportunities within the framework created by the top management. The economies of scope of the DMNC are thereby retained while economies of scale are effectively localized.

Taking an extreme flexible perspective, Hedlund (1994) criticized any structural form for being a crude categorization of hierarchy that did not account for the detailed changes that take place within companies. This is particularly relevant when managing knowledge creation at different levels of aggregation, from the individual to the corporation as a whole. Where a hierarchical structure such as the M-Form would create institutional divisions between the knowledge holders, Hedlund suggested that an N-Form structure would be more appropriate. This is a heterarchical structure that allows each division to behave entrepreneurially (Ferguson and Ferguson 2000). Hedlund defined the N-Form as an attempt to recompose the divisions of the M-Form to allow them to work together through lateral communication at middle management level and temporary groupings of joint personnel. Senior management then operates more like that of the U-Form, where it becomes involved in operational processes in order to promote cross-links between functions. The N-Form then allows the growth of knowledge in the corporation, Hedlund referring to it as economies of depth as work experience is accumulated, though this can lead to the firm being introspective and therefore vulnerable to exogenous developments. Indeed, just as Chandler states that structure should follow strategy, so Hedlund conceded that the choice between the M-Form and the N-Form depends on the nature of the firm's environment and that a compromise between the two structures may be more appropriate.

A compromise between organizational structures might be achieved by having two in place concurrently. For example, a hybrid of the M-Form and U-Form shows the M-Form decomposition of divisions in order to separate strategic and operational decision making. As before, this preserves the system of internal controls based on vertical flows of information in order to conduct efficient allocation of resources. Into this is introduced some of the functional character seen in the U-Form, such as marketing or human resources, which might, depending on the organization, be better able to serve the company as a whole (Moschandreas 1994).

If the divisions need to overcome some kind of separation problem, be it spatial or technical, then a matrix approach can be implemented. When disparate functions would benefit from being coordinated with each other, the matrix permits two lines of communication: vertically through the hierarchy, and horizontally between the functions. It is not an alternative form of corporate structure, since it can exist within an M-Form, for example, but it does have implications for the degree of decentralization and the role of managers (Ferguson and Ferguson 2000). It does not have a preset format and Kolodny (1979) considered the matrix to be a flexible coordination device, the level at which it was applied in the enterprise depending upon the nature of the information-processing demands. The most well known application of the matrix theory is the aircraft industry, where the range of technical processes requires a multidivisional M-Form; but since the divisions are working towards unified projects targets, often in an exploratory or research environment, it is necessary to establish close communication between the divisions. In application the matrix system is often transitory, tending to be shaped according to the nature of the project, with the divisions reporting to both the divisional management and the project management as necessary. It is feasible, then, that the matrix system could even be applied to a U-Form enterprise. This would then bring the functions into closer contact, particularly useful if the final product involves different production technologies. Costs rise, though, when simultaneous projects duplicate efforts being made elsewhere or when the command structure becomes too complex.

Although a variety of governance structures has been discussed it seems that each is an adaptation of the M-Form, though applied with greater flexibility. In essence, the choice of organizational structure then comes down to the U-Form or the M-Form, the M-Form being the one that seems to attract the most attention since it describes the complexity of larger corporations. Yet the two systems may not be mutually exclusive but instead view the corporation from two different perspectives. If the purpose of management is to control the activities of a company, then it could be argued that the U-Form simply describes the system of command, each level holding responsibility according to its position in the system. Thus for a large company, lower management would tend to have more operational control while higher management would tend to have more strategic control. Conversely, the purpose of the M-Form is to formalize the transfer and evaluation of data within a large or complex organization. The fact that senior management then use the data to make strategic decisions is due to their elevation in the command system rather than the official definition of their role. Indeed, the need to keep strategy and operations in alignment suggests that senior management should indeed have a close interest in operational matters, albeit through the proxy of lower managers.

Toyota's domestic Japanese structure seems to demonstrate that the M-Form and U-Form can coexist. In its *keiretsu* structure the company does not simply hold close connections with its component suppliers, the majority of its production facilities are also given autonomy to the extent that they are technically separate companies. For example, the RAV4 SUV and the Vitz/Yaris small car are assembled at a plant operated by Toyota Industries, a company that was founded before Toyota Motor and in which Toyota Motor holds just 23.51 percent of equity (Toyota 2007b). Since the two companies have been interlinked for over half a century the relationship is clearly a stable one. The result is a unique form of structure for Toyota Group's operations which sets it apart from other automotive manufacturers and so might even indicate a set of advantages peculiar to the firm. The plethora of models made at affiliated plants indicates high flexibility in production, Toyota Auto Body having 16 models on its production roster. This would suggest that Toyota Auto Body has been successful in attracting business from Toyota Motor over and above its internal *keiretsu* rivals, and therefore represents a lower cost than even an internal Toyota Motor facility. It is quite feasible that this explains the success of such concepts as *kaizen* (continuous improvement) which for affiliated companies represent a genuine incentive to attract continuing custom, rather than for internal Toyota divisions where the concept of *kaizen* must be imposed by senior management. This is not a relationship that Toyota Motor has extended to its overseas operations, where it needs to simplify operations in order to maintain close control. As a consequence, Toyota tends to produce a wide variety of niche models at home for export, while overseas operations tend to produce a limited number of models in high volumes

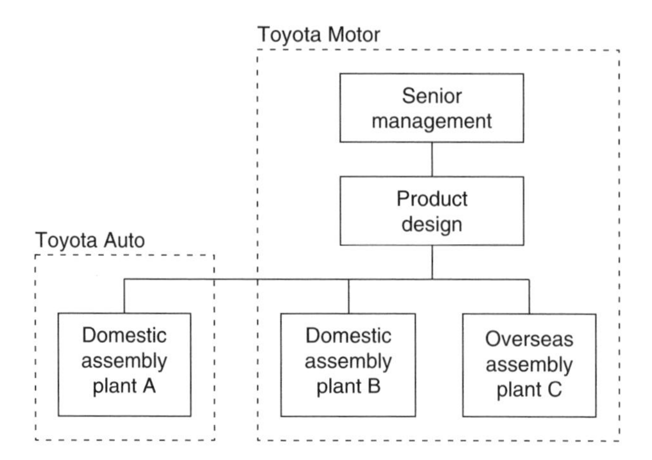

Figure 4.2 Toyota Group's domestic and international structure

for the local market. Figure 4.2 offers an abbreviated illustration of Toyota Motor's structure in Japan and overseas.

Figure 4.2 shows that Toyota is able to practice close control where necessary, while granting semi-autonomy where it can. Although the U-Form structure appears to describe a firm that cannot cope with diverse functions, the M-Form also seems to describe a firm whose senior management would be incapable of dealing with a crisis at the operational level. In practice, both views seem indefensible. The exact limit to the U-Form's effectiveness would seem to depend on the psychological capability of the CEO, yet even the smallest firm is a hierarchy of shopfloor workers, supervisors, and middle management, each exercising a mixture of operational and strategic influence to varying degrees. It would take an extraordinarily autocratic CEO to take all the operational decisions while also ignoring all suggestions related to strategy. Conversely, while the M-Form charts the flow of information in a defined format, at times of operational crisis, such as a debilitating industrial action or a crucial new model launch, it is highly unlikely that the senior management could resist the temptation to become directly involved. Indeed, given that they came to their positions of seniority after years of experience at the operational level, it would be a dereliction of duty not to participate and bring their skills to bear. While this may suggest a reversion to the U-Form in a time of upheaval when close control is required, it could be countered that the U-Form was always implicit in the command structure even while the reporting structure was in the M-Form.

The current automobile industry presents particular challenges to the M-Form structure since it is made up of technically distinct functions which are required to work closely on joint projects almost as complex as in the aircraft industry, yet they may also be geographically spread. In the following chapter I will investigate how these considerations impact on the organizational structure and governance of automobile manufacturers. With the inclusion of the prescribed economies of scale set out in Chapter 3 it will be possible to devise a paradigm for the automobile industry in terms of company size and shape.

CHAPTER 5

Constructing the automobile industry paradigm

Paradigms are an established method for viewing the advance of knowledge from a systematic viewpoint. Instead of seeing human progress as a chaotic accumulation of facts, theories, and processes, paradigms group these endeavors into communities focused on a specific principle. This creates fields of knowledge, each strictly defined and mutually exclusive. Although the concept of paradigms is well known in science where the founding principles for each area of study are explicitly stated, examples of paradigms abound in other areas of human activity. These include astronomy and astrology, music and painting, Darwinian evolution and creationism. No matter how free-thinking adherents to a paradigm might like to believe they are, in practice they are bound to the rules that govern them. Thus it is no more possible to foretell the future by landing a scientific probe on the surface of a planet than it is for a novel to be painted. Within the paradigm, the nature of the problems that can be tackled are preordained and the methods by which they can be solved are preset. Consequently, the paradigm has clarity of focus and a defined area of relevance.

Industries can be considered to be paradigmatic when they have developed within confines as strict as scientific paradigms. The parameters are set by the organizational structure and the production technology. Technical progress is then demonstrated by a process of puzzle solving, which explores the limits of the technology and brings the manufacturers closer to the theoretical output optimums. Organizational structure will arise by a similar iterative process as companies converge on a commonly held degree of vertical integration. The paradigm reaches its conclusion with a firm that can access all the benefits of vertical integration and economies of scale. In short, it takes on the ideal shape and size, a position of ultimate development that is only undermined when a new technology changes the fundamental defining principles. This prompts the development of a new paradigm which has almost nothing in common with the old one, and so presents a whole new set of opportunities for new firms.

Since car manufacturing is constrained by parameters of scale and integration it is proposed that this can be viewed as a paradigm. If this is the case then development of the industry has not been driven by progressive linear innovation for its own sake but by a process of internal problem solving. The paradigmatic view of the automobile industry demonstrates how companies must adhere to the strictures of the paradigm in terms of scale and vertical integration in order to operate at their optimal levels. This assumes that there is no quantum change in the underlying technology to cause a paradigm revolution, which in any case would have no effect on the internal characteristics of the current paradigm. To understand how progress is possible in a world governed by paradigms, it is first necessary to understand the nature of paradigms themselves.

5.1 Paradigm revolutions

Paradigms are frequently cited as providing a structuring framework for scientific research. Kuhn (1970) is the main proponent of the paradigmatic view of scientific thought. He emphasizes the revolutionary nature of scientific progress which is rooted in the sociological and methodological structure of scientific communities. Kuhn sees "normal science" as being one where research is based on previous scientific achievements that have come to be accepted by a scientific community as the foundation for future practice. He then relates this to his own concept of paradigms where scientific practice, comprising law, theory application, and instrumentation, form the basis for cohesive traditions of scientific research. For Kuhn, science itself is defined by paradigms, and within a paradigm normal science can be practiced (Chalmers 1983). Chalmers clarified this notion of normal science as "a puzzle solving activity governed by the rules of a paradigm" (ibid., 92).

According to Kuhn, normal, or paradigm-based, research is said to possess three foci: definition of a revealing class of facts, a comparison to the paradigm's predictions, and continuing articulation of the paradigm theory. He refers to a mature scientific community, a community made up of factions loyal to rival paradigms. He states that the period of normal science within the paradigm is an opportunity to pursue detailed research without having to waste resources defending the fundamentals that underpin the paradigm, these being already generally accepted. This would then lead to a paradigm taking up permanent residency within its field of research were it not for the periodic revolutions that enable human knowledge to lurch forward. It is the well-defined, self-referential nature of paradigms that means a competing paradigm will be exclusive and the transition to it must be revolutionary. Kuhn sees a dominating paradigm providing a clear set of laws governing the activity of scientists, until they lose faith with the laws and the paradigm in a revolutionary change. As Williamson (1998, 25) puts it: "it takes a theory to beat a theory." Lakatos points out that this means that revolution is extra-scientific: "Scientific change is a kind of religious change" (Lakatos 1970, 93).

The message Kuhn has for practitioners within a paradigm is to engage in detailed research but be aware that a developing mismatch between the theories of the paradigm and the outside world will necessarily entail the creation of an entire new paradigm. Thus, paradigmatic science can be said to be an organizer of theory, data collection, and testing, permitting progressively focused research without the need to reiterate accepted fundamental concepts. It is not a theoretical stance but a structured system based on a founding concept. When this concept is objective, such as a form of technology, then the paradigm describes the objective, even physical, consequences of the core technology. An industrial example of this might be the Bessemer process for the mass production of steel where the process sets limits to the production rates, material quantities, use of labor, and even plant location. Indeed, paradigms can be identified in industry wherever there is a defining technology.

5.2 Constructing the automobile industry paradigm

For the automobile industry the defining process is the Budd system of manufacturing steel unitary bodies, encapsulated within the Budd Paradigm. This process is highly capital intensive but the machinery is restricted by the nature of the material and the intended use of the product, so that the use of the technology is effectively fixed. The other processes of powertrain, R&D, and final assembly are similarly constrained so that innovation tends to be concerned with puzzle solving rather than changing the fundamental technology. As a consequence, when the four main processes are brought together the result provides a basic structure for the industry and the companies that comprise it. The degree to which a company internalizes the four main functions is decided by its relationship with the market, and since markets are becoming increasingly homogenized by the forces of globalization it can be reasoned that all companies will gravitate towards the same organizational structure. The production technology, though, has implications for economies of scale which creates a gravitational pull towards prescribed output quantities.

Economies of scale are most closely associated with physical assets since they are often based on capital investments designed for volume production. Human assets are constrained in their output and may rapidly exhibit diseconomies of scale, particularly when related to management factors due to the fixed nature of managerial capital (Levy 1985). In either case, as has been seen previously, there is a point of equilibrium where economies of scale are being exploited to the full. A firm can access these cost benefits by owning the physical assets of that process or buying from a company that does so. Lyons (1995) found that UK engineering firms tended to access economies of scale through the market mechanism, yet this would also imply additional costs related to the making of transactions. It would only be to the firm's advantage, therefore, if the marginal benefit of accessing economies of scale through the market exceeded the marginal cost of transacting in the market. Furthermore, it might be presumed that transacting in this way will always provide some access to the supplier's economies of scale.

D'Aveni and Ravenscraft (1994) investigated the effect that vertical integration had on costs when economies of scale are available internally and found that additional economies were uncovered even after controlling for the effects of economies of scale. However, the authors also found that internalization insulated the company from market pressures and so a slackening of incentives to minimize costs allowed inefficiencies to emerge. McAfee and McMillan (1995) concluded that much of this effect was due to the chance for opportunism as the company grew larger and that the hierarchical structure creates a distance between the management and the workers that have the knowledge. This results in a principal–agent stand-off: "the holder of the information exploits the bargaining power the information gives him, earning rents; in anticipation of this, the principal manipulates the outcome" (ibid., 406).

The authors see these rents as a necessary evil, as the internal "lubrication" to the mechanical friction of transactions. This, then, sets an upper size limit on the firm and reverses the conventional view that an industry is competitive when the firms within it are relatively small, the authors arguing: "thus firms are small because the industry is competitive" (ibid., 402).

This suggests that integration is not a simple matter of internalizing economies of scale enjoyed by an external party because in the process the advantage will be countered by rising costs elsewhere. The decision whether to make-or-buy depending on the transaction costs is what limits the vertical boundaries of the firm, but implicit within this is the trade-off between economies of scale in production and diseconomies of scale in the hierarchy. This points to a standard firm size and organization when three conditions are satisfied:

- The production technology underlying the quantification of economies of scale is accessible throughout the industry.
- The transaction cost considerations are common to all participating firms.
- All the human assets behave in largely the same rational manner within the same organizational structure.

In Chapter 3 I outlined how economies of scale influence firms within the automobile industry, providing a reference point towards which companies converge. In Chapter 4 it was shown that it is the rational behavior of the human assets that induces an additional cost, a transaction cost, which then forces the make-or-buy decision. This is driven by a vulnerability to specific assets which may have the potential for *ex post* advantages that accrue asymmetrically to the owner of the assets, offering them an opportunity to create a hold-up. Since no contract can account for all contingencies, if the costs of such opportunistic behavior are relatively high then the firm will integrate the activity into its hierarchical structure. Since opportunism has a human source then integration is primarily concerned with obtaining control of human assets through labor contracts, although this may then result in a principle–agent conflict instead.

There is, though, the possibility that the human actors may not act as rationally as economic theory might suppose. Williamson (1973) conceded this point in listing the three human factors that influence transaction costs: (i) bounded rationality for the human limits of knowledge; (ii) opportunism for the limits of human morality; with (iii) an "atmosphere" that acts as a sociological limit to acting upon information. While the human actors may become aware of information that points to an economically advantageous action, there may exist social mores that inhibit them from exploiting the advantage. As Williamson (1973, 317) stated: "Individuals who value independence highly may favour markets over hierarchy, while others may favour internal organization because of associational satisfactions which they derive."

This creates a problem when proposing a model for an industry, since it must accommodate those individuals who might wish to base their selection of organizational structure not on economic factors alone but also as an expression of their individuality. Yet, for economies of atmosphere to be crucial in determining organizational structure, they must be powerful enough to overcome economic factors, not just influence them. Whipp and Clark (1986, 33) noted that US corporations exhibited a predilection towards productivity over innovation, but also conceded that this was "well suited to the American market." Williamson (1996) gave some indication of the extent of atmosphere by subdividing "self-interestedness" into self-serving opportunism and frailty of motive and reason. This frailty affects only quotidian operations: "if they slip, it is a normal friction and often a matter of bemusement" (ibid., 49).

Such frailty is hardly of a magnitude that would seriously threaten the economic structuring of an organization. I view atmosphere as being of a comparable intensity, springing from personal preference. Should a participant in a business be swayed by such personal concerns over economic factors one might even wonder if this will inevitably inhibit the sustainability of the firm, given that the financial viability ultimately rests upon the management of economic costs rather than personal preference. This might even suggest that economies of atmosphere are more relevant to corporate failure, particularly if they involve cases of management that is reluctant to wholly embrace the economic argument. In any case, since individual preferences are highly variable, in sum they may be inconclusive, at least in larger firms. This is not the case for the other major human factor, opportunism, which will only vary in degree but will always have the same directional vector (i.e. a temptation to exploit the business relationship for personal gain). In proposing a general model for an industry, therefore, it is valid to focus on the technical and organizational cost factors that are generally prevalent.

Assuming that firms in an industry are faced with the same technical and organizational costs then a common type of economic foundation should become apparent. Output is prescribed by the presence of economies of scale, while vertical integration is prescribed by the presence of transaction costs resulting in a full-function structure (see Chapter 4). In the next subsection I will investigate how the two forces of transaction-cost economics and economies of scale create a paradigmatic structure for the industry and thus the prototypical automobile firm.

Paradigmatic view of the automobile industry

The paradigmatic approach can be applied to any of the major functions of automobile manufacturing depending on the distinct nature of the underlying technology. Nieuwenhuis and Wells (1997) termed one process in automobile production, that of body panel pressing, as the Budd Paradigm, and they believed that, as the process that defined the final product and the economies of scale for the industry, it therefore structured the industry (see p. 71). Later developments in powertrain, particularly in the field of electric vehicles, show that the Budd Paradigm is not one that rules over the production of vehicle propulsion systems, so it cannot be considered to encompass the entire production process, only that connected to body fabrication and painting. However, the Budd Paradigm predominates, since it is this that sets down the MES for the firm, other processes being multiplied to equalize with the production of car bodies. This implies a structure for an automobile company but it does not explicitly address the issue of vertical integration, which as has been suggested is driven by TCA. Thus, the Budd Paradigm does not of itself imply a size and shape for a prototypical automobile firm.

When the different functions of automobile production are defined by a common form of technology then the fixed costs are effectively common throughout the industry. If markets for sourcing the factors of production are generally equal for all participant companies, as they would be for a capital intensive industry such as this, then the transaction costs in procuring those resources will also be common to all companies. This means that there is an optimal structure to the extent of vertical integration for the firm, just as much as there is an optimal output for prescribed economies of scale. Considerations of scale and vertical integration therefore arrive at a fixed size and shape for a prototypical automobile manufacturer. Structural differences would only arise where there are differences in the fundamental technology, which then changes the capital costs, or else changes in the factor endowments. Since labor affects mainly final assembly it does not have an effect on the production technology in the body or powertrain fabrication processes. R&D is also more concerned with the knowledge contained within the human assets rather than labor costs per se. Labor costs therefore have the greatest influence on the internal structuring of the final assembly process.

Developments in technology have the potential to change the paradigm by affecting the economies of scale and the transaction costs. It has been suggested that new powertrain systems might change the industry, but so far this has been confined to hybrid engines which use electric motors only in an auxiliary role. Were there to be a wholesale switch to electric forms of propulsion then, although it would not necessarily affect body production, it could change the manner of the integration with powertrain production. Car firms might find that there were cost advantages to be gained from divesting themselves of their new electric powertrain production units and instead sourcing production externally.

The Budd Paradigm, as applied to BIW production, would be unaffected by changes in powertrain production technology, though a new MEPS for powertrain would have an effect on the lowest common multiple when matching the scale of the different functions. For the automobile paradigm being described in this chapter, which includes scale and the full-function model of organization, outsourcing of any function would have an impact. Indeed, this would represent a move to a new automobile paradigm – the paradigm revolution described by Kuhn. Essentially, competitive paradigms can remain side by side since both provide acceptable solutions to their practitioners in respect of the problems they define. As the new paradigm proves itself more adept at solving a wider range of problems it gathers greater support. Kuhn claimed of those that adhere to the superseded paradigm that they "are simply read out of the profession, which thereafter ignores their work" (Kuhn 1970, 19).

For automobile production, in the early twentieth century the current paradigm superseded the craft production paradigm where coach-built wooden bodies were attached to separate steel chassis. It is not that craft production has disappeared altogether, for low volume production or individual customization it is still an option, but it has been bypassed as an effective means of manufacturing standardized mass-market vehicles on a large scale. Indeed, puzzle solving can carry on in the preceding paradigm, even if not with quite the same intensity as in the new paradigm. As Kuhn stated: "mopping-up operations are what engage most scientists throughout their careers" (ibid., 24).

A paradigm is successful while the adherents continue the "puzzle solving" activities. This is the fruitful part of a paradigm, and the existence of, as yet, unexplained phenomena are viewed not as weaknesses but simply as anomalies awaiting further research. The conditions for a revolutionary change to a new paradigm are set when the adherents of the existing paradigm find that it no longer adequately matches nature and they lose faith in it (Chalmers 1983). For industry, these problems arrive in the form of exogenous shocks, but not even dramatic rises in oil prices have done anything other than shift market preferences for cars. Not only did the oil price increases have no impact on the fundamentals of the narrowly focused Budd Paradigm, but also no impact on the automobile paradigm put forward in this chapter because the powertrain function is essentially unchanged. In the meantime, these exogenous shocks have served to further motivate internal problem solving. It was one such exogenous shock that motivated Honda to design the CVCC lean-burn engine in order to meet emission regulations in the US (Demizu 2003): this represented the development of existing technology rather than a technical revolution.

Since paradigms are viewed as sociological constructs their characteristics can only be indicated, not proven. Moreover, those within the paradigm may be unwilling to recognize its existence when it might suggest their research is thereby constrained. However, paradigms aid comprehension of areas of activity since a conceptual framework can be applied, along with the methodological and predictive powers of the paradigm. Applied to an industry, this would prescribe a future structure that could be superseded only by a paradigm revolution based on a change in the fundamental technology.

Such a paradigmatic perspective on the automobile industry is only valid if the underlying technology allows it. It can then be demonstrated how this is organized according to an established approach to arrive at a prescribed structure for a paradigmatic automobile company. This implies that any car company operating with the same fundamental technology would be governed by the same paradigm and so would be drawn to the same ultimate structure. Any company within the paradigm that did not achieve the prescribed scale and structure would be deemed to be operating at a suboptimal level (i.e. operating at above minimum cost).

5.3 Vertical integration of the automobile paradigm

When a company takes more responsibility for production processes by internalizing the supply chain it is said to be increasingly vertically integrated. A vertically integrated car manufacturer is one that has internalized all the major functions of automobile production. Car manufacturers have often explored the limits of vertical integration; for example, Henry Ford built a steel mill for his new factory at River Rouge (Rae 1984). However, the purpose of this book is not to include every last detail of automobile manufacturing but to investigate the main structuring areas. The perspective taken of car firms in their generic form is therefore in accordance with the full-function model previously put forward (see Chapter 4).

The company boundary of a typical full-function automobile company with its four modules of production is shown in Figure 5.1. This is a functional schematic of the vertical integration of the firm; it does not imply actual plants or their geographic positioning. The design function instructs the powertrain and body functions, the output from these feeding into the final assembly function. The final products are released to the market, which is external to the functional operations of the company. Information from the market is fed back to the company's design function by way of a marketing capability which is part of the R&D activity.

In theory, any of the major functions can be external, including design and final assembly. However, from a historical perspective, the default position for a car company, before it vertically integrates, has been to

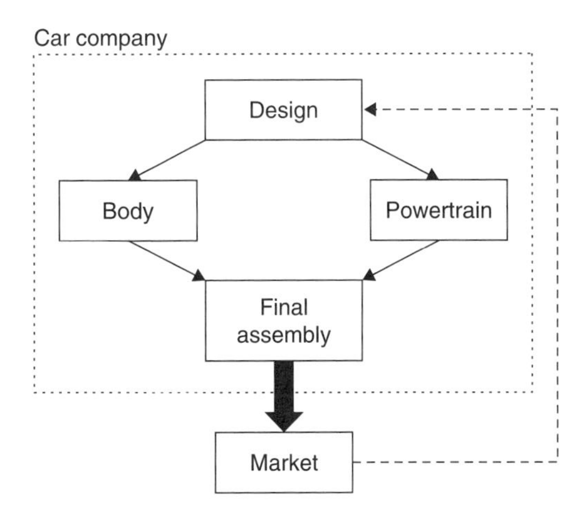

Figure 5.1 Typical full-function automobile producer

commence with design and final assembly functions. In this way the vehicles are conceived and put together by the new firm, sourcing all components externally. Rare examples of production existing before the R&D capability include Tata of India, which had its Indica small car designed by IDEA in Italy, although Tata had control over the parameters of the design process (SID 2006). However, as will be demonstrated, most mass-market firms start with the design and final assembly functions at least, and then integrate powertrain and body as the business develops.

Since such an integrated structure takes time to develop and become established it is necessary to research archive data. The fundamental technicalities of automobile production were first introduced at the beginning of the twentieth century with the first experiments with flow production systems and the rise of mass production. This brought vastly improved assembly methods and efficiency benefits to the existing forging and casting techniques used for heavy components like powertrain. Finally, body production was revolutionized by the Budd system of manufacturing load-bearing welded steel bodies that brought clear constraints to the style of vehicles that could be produced and indicated what the economies of scale might be. Once the technicalities of production were defined it was possible for a stable organizational structure to emerge. The process by which vertical integration occurs is not a static one and it is as possible for firms to divest themselves of their functions as it was to integrate them in the first place. By this reasoning, if functions remain internalized then, rationally, the transaction costs must continue to be minimized. The purpose of this section of the book is to build a representation of the core functions of automobile production as they have become established over time, whilst also noting how they might have been adjusted in detail to take account of the contemporary conditions.

Defining the functions

As a multistage process, automobile production involves inputs from a myriad of component suppliers. Wells and Rawlinson (1994) conceded that there are difficulties in identifying the limits of the supply industry, the numbers of firms, or even whether they can be considered automotive or not. However, for the purposes of this book, it is important to define how these firms relate to the automobile assemblers so that their role outside the vertically integrated full-function structure can be delineated.

Even for a full-function firm operating in the four core areas there will be component suppliers contributing at the subfunctionary level. These range from producers of small individual parts to larger subassemblies that are built up from a collection of components before the completed unit is fed into the main assembly line. The illustration of a full-function automobile company (Figure 5.1) has omitted the substantial role played by this supply industry. Firms in the supply industry can be categorized into three types according to the nature of their output and its use (Odaka *et al.* 1988):

1. OE (original equipment) for installation in complete vehicles during final assembly.
2. RP (replacement parts) or SP (spare parts) to replace worn out parts.
3. Reconditioned parts, as low cost substitutes for RPs and SPs.

Since OEs are made to the design or requirements of the vehicle assembler, unlike the other two categories, only supplies of OEs are of direct interest here. A noticeable feature of this ancillary industry is its tiered structure of primary and secondary suppliers, the primary or tier 1 suppliers often being subassemblers using parts manufactured by the secondary or tier 2 suppliers, although some secondary parts may be procured directly by the vehicle assembler. There is a tendency in Japan for suppliers to be tied exclusively to one producer, as exemplified by the *keiretsu* structure. This may also include a degree of supplier selection within an obligation model of the supplier–buyer relationship wherein a preferred supplier is groomed for a high profile role. The supplier retains semi-autonomous control over production, while the buyer can influence product specification, R&D, and investment decisions (Morris and Imrie 1992). This is analogous to the M-Form of internalized organizational structure in that the supplier has operational control, while the buyer sets the overall strategy and capital investment commitments.

In Europe the relationship tends to be more polarized, suppliers being either majority owned or independent of the car producer (Wells and Rawlinson 1994). When the supplier is independent this can lead to an adversarial relationship with cost reduction as the battleground. This operates through a tendering system that promotes a divided supply industry that is unable to make long term commitments to R&D or economies of scale in production. Smitka (1991) noted that Chrysler operated a policy of short-term contracts with suppliers that kept costs down through intense competition, but inhibited long-term investment in the supply industry and so stifled development.

Despite this, Carr (1993) found little correlation between UK supplier profitability and production volume, instead sourcing the decline in the supply industry to the 40 percent rise in the currency exchange rate in 1979, which forced domestic suppliers to hold prices for four years in order to resist foreign competition. This left little room for reinvestment, although the study found that GKN was able to maintain world-leading R&D expenditure up to 1983. This suggests that the connection between price tendering and low investment is not a direct one. Moreover, a competitive market is not described by the price mechanism alone but also by that of product features, so it is unlikely that price tendering represents one-dimensional contracts but is more likely to be one aspect of multidimensional contracts. It would seem, therefore, that the problem with the adversarial relationship is that during periods of change (such as exogenous economic shocks) low switching costs mean that there is no reason for the buyer to support the supplier, thereby ending the relationship. It is notable that part of Nissan's recovery has been predicated upon an imposed 20 percent reduction in purchasing costs spread over three years and a 50 percent reduction in the number of suppliers (Donnelly *et al.* 2005).

The varying levels of integration between suppliers and car producers are illuminated by the range of relevant factors stated by Odaka *et al.* (1988). This includes technological distinctiveness of the part, production technology, demand, and economies of scale. The core functions of the full-function car producer are technologically separate but closely tied in terms of the production system. For instance, the body panels are used solely by that assembler. This is not necessarily the case with other components, which may use an unrelated technology and may be shared with other car producers. Rhys (1972) reported that Lucas had a virtual monopoly in the

UK for lamps, windscreen motors, dynamos, and ignition coils. Toyota also recognized that some components were of a generic technology that could be purchased from a variety of manufacturers, while other components were of a specialized nature and therefore required Toyota specification and "teaching," secured by a close capital or financial relationship. The procurement category of "special factory purchasing" requires "distinctively special facilities," leading to possible financial ties in the future (Lamming 1993, 24).

The main consideration in this relationship between supplier and vehicle assembler is governed by TCA. For generic components the car company has purchasing power and can switch suppliers for a variety of reasons, be they cost, delivery, quality, and so on. As Odaka *et al.* (1988) have pointed out, the supplier is operating at a different technical, productive, and scale level to the car producer. For this reason, a fuel system supplier is subfunctional to the powertrain production which a gearbox or engine supplier is not.

For each function it has been seen that there is a critical MEPS and that the relationship with the supplier is functionary when the supplier is responsible for this scale of output. This is a proxy for asset specificity since, although physical assets may not be specialized, their workload is to the degree that without the contract to supply they would be redundant. Monteverde and Teece (1982) also discussed the crucial role played by the human assets where the supplier is engaged in research in support of the car manufacturers own R&D, arguing, like Klein (1988), that this will lead to vertical integration. As will be seen in the following subsections, when suppliers occupy a complete function such as powertrain production then they tend to become internalized into the automobile manufacturers' structure.

Integration with powertrain

External supply of powertrains was common in the early years of the automobile industry. Thoms and Donnelly (1985) focused on the cradle of the British automotive industry, Coventry. White & Poppe was the main engine supplier in the city, and in 1906 the company was supplying 15 car manufacturers. In 1913 engine output reached 2,000 units but it progressively lost business to car companies seeking to vertically integrate powertrain production. Daimler, for example, produced its own engines and in 1908 planned for an additional output of 1,000 engines, of which 400 to 600 were destined for Rover. White & Poppe was forced to invest in specialized capital as technology progressed, but this created an asset specificity problem with Singer extracting credit terms from the company in 1912. The directors at Standard were aware of the power they had over White & Poppe, agreeing that if "conditions warrant it, deliveries can be suspended at any time according to requirements" (ibid., 88).

Engine maker Hotchkiss & Cie was closely involved in supplying Morris, and in the two years after being vertically integrated into Morris output quadrupled (Thoms and Donnelly 1985). In the US, since a sufficient supply base had yet to emerge in the new industry, Henry Ford founded the Ford Manufacturing Company in 1905 to ensure a reliable supply of engines to required specifications and prices, which was then consolidated with the Ford Motor Company a short time later in 1907 (Hounshell 1984, 220).

Since most of the vertical integration with powertrain suppliers occurred in the early years of the industry, there are few instances of it occurring in the modern era, even if that is taken back to 1945. I am not aware of any significant automobile firms divesting themselves of their powertrain functions. Later entrants, such as Honda, began as an engine manufacturer in 1946 and then branched out into motorcycles in 1949, followed by cars in 1967 (Honda 2006). In Korea, Kia began engine production in 1973, a year before it began production of its first car (Kia 2007). The Chinese manufacturer Chery was established in 1997 and began engine R&D at the same time, designing the ACTECO engine family in partnership with engineering consultants AVL of Austria; annual engine production capacity is 400,000 (Chery 2007).

External engine suppliers operate now on a contractual basis with limited output. For example, VM Motori of Italy supplies diesel engines to three major groups (DaimlerChrysler, LDV, and BMC) with licensed manufacturing granted to Hyundai in Korea (VM Motori 2006). It is quite common, however, for powertrains to be supplied from one division to another within a single group, such as VW's W12 engine (in its basic form) to Bentley Motors.

Integrating body production: the case of GM and Fisher Body

Body production is an interesting example of integration because the coachbuilding industry preceded the emergence of the automobile industry, and so the early automobile manufacturers were able to source bodies externally from an established market. Subsequent innovations in body production that led to the load-bearing steel body created a revolutionary upheaval in the industry which was addressed by internalizing the body production function into the automobile manufacturers. In the case of GM and its body supplier, Fisher Body, this has become a celebrated example of internalization which resulted in the application of the M-Form of organizational structure. It also brings a critical perspective to concepts of economies of atmosphere and geographic proximity of manufacturing facilities. Subsequently, most automobile manufacturers have retained this function internally, although there has been some limited growth in contract manufacturing of steel bodies.

In the craft era of automobile production the vehicle bodies were commonly manufactured by large numbers of external suppliers able to produce in small volumes to customers' specific requirements. Automobile firms were, in the main, assemblers of outsourced components, the ultimate shape and purpose of the final vehicle being defined by the coachbuilder. Even when the automobile firms reached mass production levels of output, the separate body could be manufactured externally in a plethora of different body styles, albeit to automobile firm specifications rather than that of the coachbuilder.

The emergence of the Budd Paradigm changed that because the system resulted in continuous high volume production of standardized bodies which defined both the shape and purpose of the vehicle. According to transaction cost theory, the high asset specificity and high uncertainty during the changeover to the Budd system would have culminated in vertical integration (David and Han 2004). Accordingly, the absorption of Fisher Body by GM from 1919 to 1926 was one of the most notable cases of vertical integration in the automobile industry and an example of Chandler's view that structure followed strategy. Although this took place during the early years of GM's formation into an M-Form style organization, the essentials of the Budd system were being put in place, and they have not fundamentally changed since that time. The steel panels were stamped out and welded in a manner that cannot change due to the nature of the material, although there has been some refinement of the process. For this reason the fundamentals of the story still shed light on the industry today, Freeland (2000) describing the case as "paradigmatic" and Klein (2000, 107) stating that "we can learn a great deal about the economics of contractual arrangements by studying the conditions under which the Fisher–GM contract failed."

The situation in 1919 was that the vast majority of automobile bodies made for GM were of the open type – only 10 percent were closed body styles – yet by 1927 the proportion of closed body styles had risen to 85 percent (Sloan 1986, 152) due to the enabling technology of the Budd Paradigm. This was pioneered by the Essex division which reduced closed body prices such that between 1924 and 1926 prices of the Essex coach converged with, and then undercut, the open version (ibid., 159; Georgano 2000). As Sloan pointed out, with Ford pursuing a static policy of minimal product development with the open body Model T the demand in the market for the new closed automobile was taken up by competitor automobile producers, Chevrolet's share of this new market rising from 40 percent in 1924 to 73 percent in 1926.

Fisher Body was a dominant force within the coach-building industry and GM was vulnerable to its supply. GM had tried to secure the relationship with the contractual approach, taking the precaution of obtaining a 60 percent holding in Fisher Body. This was enough to preclude the Fisher brothers, as owners of the voting stock, from merging the company with other manufacturers, particularly Willys-Overland in Cleveland. However, it did not give GM operational or strategic control of Fisher Body and supply of bodies to other vehicle manufacturers continued (Freeland 2000). GM had also entered into a ten-year contractual agreement sourcing bodies from Fisher Body for a payment of cost plus 17.6 percent. According to Klein (2000) this removed from Fisher Body the motivation to reduce transport costs, since any additional costs were covered by the contract, and indeed attracted increased profits. This was particularly relevant to GM's plant at Flint which was some 57 miles distant from Fisher Body's Detroit plant. However, the Fisher brothers were reluctant to invest in a plant in close proximity to Flint, and this constituted an apparent hold-up. Yet Klein (1988) argued that specific physical assets do not have to be owned by the supplier: the buyer can purchase them and employ the supplier to use them. Coase (2000, 29) takes a similar line, pointing to the funding by GM of a previous Fisher Body facility in order to supply a Chevrolet assembly plant: "it is not without interest that it was Fred Fisher who suggested that the body plants be built on Chevrolet property."

Freeland (2000) found that the Fisher brothers were reluctant to set up a new site at Flint, not because they were averse to making the investment but because it would have made production at their Detroit plant less economically viable. It was much the largest Fisher Body plant, twice as big as the next in size, and it enjoyed considerable economies of scale as it supplied important customers in the immediate locality. Even if GM had made the investment in a Fisher Body plant at Flint, it would have forced the Fisher brothers to restructure the Detroit plant to reduce its capacity and render production there less economic.

This positioning of the plant also demonstrates that geography was already a significant factor in the industry. Indeed, with the technical principles of the industry established, proximity of certain facilities would be a constant throughout the future development of the industry. Although the absolute costs of transport have changed over the period since the Fisher Body episode, as a cost relative to production costs and competitor costs it is as relevant now as it was then. GM's plant was much further from Detroit than its rivals and with the rising popularity of the closed steel body transport costs would have come to dominate total body production costs. This might have been alleviated if bodies were generic automotive components since GM would not have been dependent on the distant plant. Unfortunately for GM, load-bearing bodies are the least generic because they define the vehicle type. Although other body suppliers were available there would still have been switching costs due to the specialized nature of the bodies. It was also not yet clear how far the market would shift towards closed steel bodies and, with the future of GM in the automobile market at stake, it was vital for GM to be able to control its supply. As Walker and Weber (1987, 590) stated: "as uncertainty rises regarding a buyer's future requirements and, correspondingly, regarding potential adjustment costs for suppliers, contracts become very difficult to write. Producing a component in-house consequently becomes more attractive."

If Fisher Body's output had been wholly dependent upon GM then GM would not have been in conflict with its supplier. Only by being in a symbiotic relationship would a relatively costless, self-regulating relationship have been possible instead of vertical integration. As it was, vertical integration was the short-term solution to the external problems of body supply within the new Budd Paradigm. According to Klein, Fisher Body had found itself in a position where it could hold up GM, forcing it to purchase the remaining 40 percent of the supplier at a premium of over 50 percent of the market value (Klein 2000, 124).

Klein speculates that the two companies had always been aware that their contract was "incomplete" but were able to rely on the self-enforcement mechanism of their reputational capital. This was changed by the unforeseen popularity of the closed body which presented a critical opportunity for GM to overtake Ford, yet this new development also gave unprecedented leverage to Fisher Body over GM. The original contract had been formulated to prevent GM holding up Fisher Body but now GM's only escape before the expiration of the contract in 1929 would have been by voluntary bankruptcy. The stability afforded by the 1919 arrangement was inflicting "rigidity costs" that could only subsequently be avoided by vertical integration (ibid., 130).

Lieberman (1991) found empirical support for demand variability or uncertainty being a determinant of vertical integration, although David and Han (2004) have found less convincing evidence. In a stable industry,

GM and Fisher Body might have sustained a long standing relationship. Coase (2000) denies that a hold-up occurred because he cannot conceive of a circumstance when it would have been in the Fishers' interest to cause one. The brothers were, in any case, employed by GM by this stage and so it would have been injurious to their own careers to have acted for Fisher Body against GM. However, a hold-up does not have to culminate in a complete cessation of supply: it might simply involve resistance to comply fully with the wishes of the buyer. This would certainly fit the predicament GM found itself in when rivals in Detroit were being favored by the close proximity of the local Fisher Body plant.

A possible alternative for GM might have been to renegotiate the contract, but this would have perpetuated the problem of rigidity in an industry that, for GM and its annual model-change policy, required flexible responses until the scale demands of the new technology had been settled. While economies of scale in body production remained an externality to it, GM would have had no control over it. Although Lieberman (1991) found that the quantities of product flowing between the contracting parties had no significant impact on the tendency to integrate this did not recognize the critical role played by economies of scale. Once the quantities involved in the deal are within striking distance of achieving significant economies of scale then they become internal to the relationship and it is in the interests of the buyer to have some control. Coles and Hesterly (1998, 327) were of the opinion that economies of scale were crucial to vertical integration when the firm could achieve these economies internally just as much as a supplier could. Fisher Body was in a position where it could enjoy economies of scale in the new production system by supplying a diverse number of car assemblers. Yet this describes a typical supplier–buyer relationship, GM only needing to internalize Fisher Body if it could better exploit the economies of scale. Since forcing Fisher Body to invest in the Flint plant would have, in the short term at least, reduced the access to economies of scale, there was no reason on these grounds to integrate vertically.

As owner of Fisher Body, GM could gain control of the new technology and avoid the costs of renegotiating contracts at frequent intervals. Although Williamson (1975) chose not to emphasize it, frequency is one of the three dimensions of the transaction cost approach. Klein agrees, concluding that the affair with Fisher Body was not concentrated on asset specificity alone – but if there is a need for flexibility when asset specificity is high and reputational capital is low then internalization protects the manufacturer. However, Coase (2000) is of the opinion that the Fisher brothers would not have wished to risk their reputation because they were supplying bodies widely throughout the industry.

There is some possibility that Williamson's concept of "atmosphere," or "personal inclination," might have some relevance here. Freeland (2000, 42) notes a reluctance on the part of the Fisher brothers to relinquish control of their family firm: "accustomed to the autonomy of running their own firm, the Fishers made it clear that they did not wish to sell their business to another company and to stay on as employees."

It would be tempting to categorize this as personal preference, and therefore attributable to atmosphere, but the brothers had started the company as an innovative manufacturer of automobile bodies, without reference to the preceding generation of carriage makers, and by 1916 the company was the largest body manufacturer in the industry. The Fisher brothers had also started the Hinckley Motor Company around 1914 to manufacture engines, initially for trucks bound for the European war (Coase 2000). In 1919 the brothers were formulating a plan to enter the automobile production business themselves, exploring a $10 million (£2.3 million) joint venture with Willys-Overland. Far from being held together by atmosphere, the siblings appeared motivated by a desire to exploit the economies of scope in their managerial and entrepreneurial skills, culminating in an automotive group to rival GM. Since GM lacked the knowledge the brothers had in body production it was vital to secure the brothers' services during a period of uncertainty: it cost GM $27.6 million (£6.3 million) for the first 60 percent holding and another $136 million (£30.7 million) in stock for the remaining 40 percent (Freeland 2000). Clearly, the brothers were not defending their family firm for reasons of atmosphere but because they believed they had a successful business model, and one with a quantifiable price.

The conclusion of the Fisher Body episode was part of the implementation of the M-Form organizational structure at GM. The senior management at GM had needed strategic control over Fisher Body and this then became a constituent part of GM's M-Form of structure. The critical component was not the cost of the physical assets at Fisher Body but the costs of the uncertainty derived from the variability in the market. Internalizing Fisher Body allowed GM to make labor contracts directly with the management, initially comprising the Fisher brothers themselves, thereby permitting greater flexibility at a time when the new production technology had yet to settle (Klein 1988). Indeed, the day after Fisher Body was acquired by GM, a $5 million (£1.3 million) investment in a body production facility at Flint was announced, while the Detroit Fisher Body plant was subsequently closed. The internalization strategy set the tone for the industry as a whole and GM's vertical integration introduced an industry wide trend (Klein 2000).

Integration with vehicle design

The ability to research and design future products lies at the very heart of any corporation. R&D is the source of all product innovation and so its vertical integration within the firm allows it to come into contact with all parts of the firm (Armour and Teece 1980). It might be argued that these contacts could be contracted for, but innovation is opportunism at its most unpredictable and fertile. Furthermore, integrating it with existing production facilities provides a focus for the R&D activity. Coles and Hersterly (1998) considered the possibility that research can make unexpected leaps forward, rendering existing technology obsolete. Should the research be conducted by an external organization then it creates an asset specificity problem: "new development in technology could render the methods used by the contractor obsolete before the terms of the contract expire. The contractor, however, may be unwilling to incur the additional costs of the new technology before the existing contract expires" (ibid., 326).

The design aspect of the full-function model is the one least likely to be external in the first place, therefore least likely to be a target of subsequent vertical integration. This is because the design function is the origin of the ultimate product, the other functions serving to bring the product into being. For this reason, automobile companies that operate without their own design capability tend to owe their existence to national and political strategies rather than entrepreneurial innovation. Many developing nations have attempted to industrialize by inviting foreign firms to enter their markets using the mechanism of joint ventures with local firms involved in final assembly. The local automobile firm may be wholly owned and a stand-alone assembly facility, but it is dependent on the foreign firm for the product designs.

Countries where this has been commonly practiced include Iran, India, and China. Iran Khodro, the national automotive champion, started out manufacturing British Hillman Hunters as the Paykan from 1967 under license from the Rootes Group, continuing under Chrysler and then Peugeot from 1978. Iran Khodro's attempts at developing the Paykan were not far reaching, although it did manage to fit the Peugeot 405 body onto the Paykan platform to create the 405 RD (rear-wheel drive) for reasons not specified (Iran Khodro 2005). India went in a similar direction, manufacturing the 1948 Morris Oxford as the Hindustan Ambassador, virtually unchanged up to the present day. Ironically, the few that are now imported into the UK have to undergo extensive modifications to meet current legislation.

China has used car assembly as a device for technology transfer. Foreign firms are limited to maximum shares in joint ventures of 50 percent, the only foreign majority-owned plant being a Honda factory manufacturing for export. In partnership with foreign firms the largest Chinese firms have achieved high levels of output, 911,748 units in the case of SAIC in 2005; but it has only been recently that indigenously designed and produced cars have made a significant impact on the market. Even so, they have yet to reach globally recognized standards, local firm Chery having to postpone entry into the US market by two years until 2009 due to difficulties in overcoming design challenges (*Automotive News* 2006c). The company is also dependent on numerous external consultancies, such as Lotus of the UK and AVL of Austria. This demonstrates that lack of geographic proximity is no impediment to the supply of R&D, although consultancy companies usually operate R&D branches in target countries.

There may be some notion that the economics of atmosphere has a role to play in the promotion of R&D capability by national governments. This can include the elevation of car firms to the status of national champions in order that they might be catalysts for wider industrialization. The implication of the political involvement in structuring the industry in China and Iran is that car companies that do not have control of product design are in a potentially unstable condition should that support fall away. The underlying instability, though, and hence the fundamental structuring factors in these cases, are based on the economic concepts that deal with the treatment of assets. R&D is defined by human asset specificity and advantages resulting from innovation will accrue to the company that owns the labor contracts of the engineers. R&D consultant suppliers therefore work on a contract by contract basis in order to control the succession of innovations. This increases transaction costs for the producers, so governments compensate for this by providing long-term support.

The efforts made by governments to promote R&D within domestic companies may be an effective policy. Zenger (1994) noted that engineers not only bring knowledge with them to their jobs but they also accumulate knowledge through work experience which makes them attractive to new employers. In this way the knowledge can become disseminated throughout the industry. I have also found that R&D teams do not need to be excessively large. It is difficult for any company to monitor and provide incentives for knowledge, but it costs less to do this in small teams because managers are in closer contact with the engineers – they may even be among their number – so have a better understanding of performance. It could also be suggested that since achievement is more readily identified in small teams it is easier for engineers to gain recognition and status. Zenger states that "small firms will attract individuals with superior talent and ideas, and will motivate higher effort than large firms" (ibid., 713).

Thus, smaller R&D departments will recruit better engineers and then compound the advantage by having them work more efficiently, though this may mean the firm losing some access to economies of scope. Large firms, as noted previously, can replicate the small-firm advantage by fragmenting into small teams, buying up small companies, or hiring small teams as outside consultants, while continuing to access economies of scope. This presupposes that contracts can be written by large companies for those engineers that are indistinguishable from the contracts that small companies could write, yet such a condition must be entirely theoretical since, if the two contracts are inseparable, then the firms must, de facto, be the same. If a firm is defined by its contractual relationships, as put forward by the theory of TCA, then only a small firm can write a small-firm employment contract.

In Chapter 3 I have demonstrated that economies of scale for all the functions in automobile production are high, but not necessarily inaccessible. The output of the functions can be brought within reasonable synchronization of each other, and in this chapter I have shown how the functions can be internalized. The next section will detail the model of an automobile company according to the considerations of scale and structure.

5.4 Resultant size and shape

Having established the organizational structure of the full-function model through the study of transaction costs it is now possible to quantify the dimensions from the earlier investigation of the economies of scale for each function or productive unit (Whipp and Clark 1986). This then results in the proposed prototypical full-function automobile company model which is at the basis of the automobile industry paradigm. Figure 5.2 shows the four functions with the MEPS for each, as in the case of the prototypical automobile company outlined in Chapter 3. Drawing on Figure 5.1, the design function is termed R&D and comprises 1,200 personnel working on five model programs based on three platforms with an intended output of 600,000 units a year. Body production is now expressed as BIW, in deference to the Budd Paradigm, with an MEPS of 500,000 units a year. The powertrain function is shown with an MEPS of 1 million units a year, while the final assembly function has been allotted an MEPS of 350,000 units a year.

Figure 5.2 shows that there would be a great deal of mismatch in output capacity between the various functions if the firm were constructed in this way. The capacity limit in final assembly creates a constriction in the

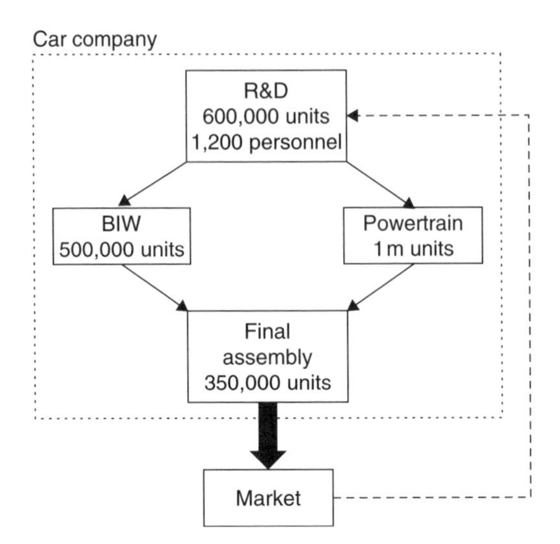

Figure 5.2 Full-function model quantified

production flow that forces the preceding functions to operate well below the scale shown. However, as the LAC curves for each function depicted in Chapter 3 show, there is a range of plant sizes which may not be operating at the optimum, though they are within the relatively flat region of the LAC curve. Taking the MEPS of one process as the anchor point, the other processes can be adjusted accordingly. If this were BIW, for example, R&D would be reduced slightly in size (perhaps 1,000 personnel working on two platforms and four models), powertrain would be half the size (500,000 units a year), while final assembly would be reduced to a capacity of two plants with 250,000 units a year each. As Chapter 3 showed, there are examples of these plant sizes within the industry.

I take the view, however, that R&D MEPS is the anchor point for the full-function automobile manufacturer because beyond that point it is inflicted with diseconomies of scale. For the prototypical automobile company shown in Figure 5.3, the MES_P is considered to occur at around 600,000 units of output a year, which is aligned with the MEPS for R&D, though it will require adjustment in the other functions to match it. This would involve BIW producing above the MEPS, powertrain well below it (but within the flat range of its LAC curve), and final assembly made up of two plants with output of 300,000 units a year. As is now standard in the industry, each of these functions would make use of flexible production to manufacture multiple models, usually variants of one another.

Since the final assembly process has been duplicated it is possible that the output split between the duplicate plants could occur anywhere within

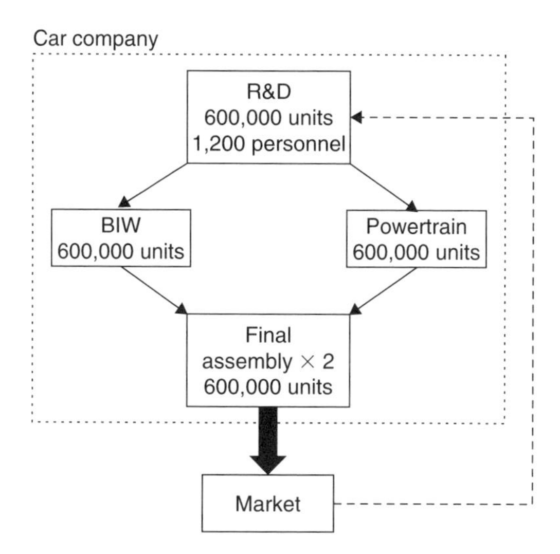

Figure 5.3 Automobile industry paradigm

the flat range of the relevant LAC, thus the total output could comprise Plant A with annual output of 350,000 units and Plant B with 250,000 units, or even three plants with output of 200,000 units a year. The extent that capacity would be allocated between the plants would depend on company specific factors, for example the larger plant might be a later addition to the company. The company might also have to calculate the degree to which higher unit costs at the smaller plant could be compensated for by lower unit costs at the larger plant.

The simplicity of the paradigm suggests that a simple form of governance structure could be applied, the most basic being the U-Form. Historically, most automobile companies have started with this structure, Henry Ford's autocratic control over the company he founded being a particular example. GM under Alfred Sloan then developed the M-Form structure in order to manage the multidivisional strategy. The focus of this book is directed more at the strategic matters of size, in terms of access to economies of scale and functional internalization. Nevertheless, governance structure is relevant in demonstrating how a paradigmatic automobile firm can manage its functions in a dynamic commercial environment. M-Form and U-Form governance structures will therefore be evaluated where appropriate. In the following subsection of this chapter I will investigate the empirical evidence collected during the fieldwork phase of this study from elite interview information on the structuring considerations within the industry. This will then feed into the next chapter which will examine the various ways that suboptimal firms can approximate to the paradigm.

Vertical integration and the extended U-Form

The fieldwork conducted for this book involved senior executives throughout the industry in the UK and Japan. Interviewees were asked about functional control and the degree to which the companies had control over their core functions. All companies involved were insistent that it was this that gave them a right to a place in the industry. R&D was seen as particularly crucial, and companies were keen to show that where inputs came from external parties this was in a supportive role and not one that impinged on their design authority. Internalization of R&D by owning the engineers' employment contracts was seen as necessary because this function is the origin of the future stream of new products. Once these products were in production, internalization of the manufacturing functions through ownership of the physical assets was seen as necessary in order to control output.

Interview data gathered at Honda showed that it placed a high priority on control of its technical assets, the latest example of this being the company's internalization of fuel cell technology after the early partnership with Ballard of Canada. A development engineer described the fuel cell research and how it progressed from a joint venture with Ballard to a sole Honda project. However, a very senior executive denied that there was any preset sequence to the relationship, only that in its ultimate form it should be wholly under Honda's control. He emphasized the cost perspective, namely that if Honda could bring in enough revenue to sustain itself then it could develop wholly owned technologies for the future and so further strengthen the potential of the company.

It was quite apparent to me that Honda maintained a strong integrated structure, despite being spread across the continents. An executive based in the UK argued that a core competitive advantage for Honda was its integration of company functions, paying constant attention to designing for production. He contrasted this with engineering consultancies that housed "excellent" skills but did not have the capability to see proposals through to production. This reliance on its own capabilities was at the very core of the company and was said to be the main strategic force. When the senior management set the general direction for the company this then allowed workers to have a degree of autonomy within the framework. Many commentators discussed the semi-autonomous nature of Honda's global divisions and this seemed to suggest an M-Form of organizational structure. The very senior Japanese executive went one step further and claimed that it was the policy of the company to encourage each major region to have full-functional capability in "development, production, sales and marketing," a point echoing the managerial theory of Bartlett and Ghoshal (1993).

A feature of Honda is that some activities have taken it outside of the automobile related full-function structure, such as motorcycle manufacturing, jet engines, and robots. This might indicate an M-Form structure, and since there was considered to be a derived technical benefit for the more orthodox automobile side of the business, this hinted at Hill's CM-Form of organizational structure. Jet engines and robots were looked on as a testing ground for product innovations, to be incorporated in automobiles at a later date. Company representatives conceded that none of the technology from ASIMO was used intact in other applications, but the research was fertile ground for additional ideas. Honda encouraged engineers to transfer between different areas of the company and gain wide experience. At the very least, it was realized that the publicity generated by such projects was a highly cost effective way of recruiting engineers of the best standard without having to offer high salaries.

However, not all functions were treated equally and interviewees reported that ultimate R&D authority lay in Japan, even though overseas locations had R&D centers. One senior engineer who had spent some years in the United States pointed out that the American head of R&D in the US had spent three years in Japan learning the Honda methods. This knowledge was then transferred to the US operation on his return in order to homogenize Honda's R&D capabilities. While production processes appeared to have more autonomy, it could be argued that they operated closely to Honda standards. Thus it would appear that Honda's organizational structure was a variation of the U-Form, with foreign extensions being allowed operational flexibility, rather than an M-Form where the foreign functions would have operational control. There was certainly no notion of the managerial theory of the firm proposed by Bartlett and Ghoshal (1993) with regard to the management of international operations.

There was some sense that a matrix format was also emerging at Honda to draw different model programs closer and exploit economies of scope. As recently as 1992, the CRX sports hatchback model program had used a dedicated design team that did not share its development with other teams. Conversely, it was pointed out that the technologies developed for the NSX supercar permeated throughout the rest of the range, the advances in lightweight parts contributing to the high standard of even the new diesel engine.

The company had achieved similar economies of scope with this "trickle down" effect before, an executive describing how motorcycle racing had been used to set the technical goal of the company.

Despite the increasingly close working relationship between projects there was a belief that there was further to go. A senior Japanese manager in the UK believed that motorcycle and car divisions needed to work more closely, perhaps to create a low-cost entry level car to bridge the gap between the two product types. He also believed that the economies of scope brought an advantage in production when expanding abroad because it could pilot production in a new location using motorcycle manufacturing. Other company representatives also mentioned the utility of gaining experience in motorcycle manufacturing before moving on to automobile production. However, only America seems to present any evidence for the motorcycle business being used as the corporate spearhead, and even at the time that the motorcycle business was being established in 1979 the planning for the automobile facility was already well advanced (Honda 2006).

The opening of new plants may blur the distinction between the M-Form and U-Form. Using one production system to test the local environment before expanding into another system suggests the M-Form, evaluating the data before sanctioning further allocation of resources. However, since the automobile plant had been planned before the motorcycle plant had been fully established, this suggests a more centralized governance structure. It seems that Honda was operating a domestic U-Form with international extensions. These extensions were allowed limited autonomy only where it was necessary to account for local differences, such as marketing and product adaptation, but these adaptations were targeted responses and operational control would be too general a description. As was unanimously agreed amongst the interviewees, the company put R&D at the core of the company, which was kept in close contact with the needs of the global markets. This strategy was further emphasized by having all CEOs promoted from the R&D function.

Research conducted at other automobile manufacturers supported the view that the U-Form prevailed. The divisional companies under Ford and VW, such as Jaguar and Bentley, had elements of both strategic and operational control such that their own governance was reminiscent of a unified U-Form structure. This again emphasized that control of entire product programs is crucial, the parent being used as a supplementary resource. This is remarkable given that the parent companies Ford and VW were also vehicle manufacturers, and of a much larger stature. Most remarkably, it was found that divisional companies could leverage their association with their larger parents to extract favorable terms from component manufacturers. In return, the component manufacturers benefited from greater security in the relationship with the parent vehicle manufacturer. The benefits of the organizational structure therefore appeared to favor the subsidiary company by indicating a system of hidden subsidies. This might suggest the internal capital market of the M-Form, yet so dependent are the divisions on these parental indulgences that it appears that the funds are not being made available internally to overcome imperfections in the external capital market but to mask the inability of the divisions to raise capital independently. It was even reported that there were occasions when the divisional firms were able to use their new membership of a larger industrial group to extract more advantageous contracts from their suppliers. This was despite the fact that the component requirements were unchanged; it was simply that being part of a larger group gave the divisional firms greater negotiating power. Where the parent groups exercised strategic control it tended to be at a very basic level, such as the setting of long-run output targets or the basic vehicle architecture. For example, Bentley uses a VW platform for its GT range while Jaguar has been dependent upon Ford for two platforms and a final assembly plant.

Throughout the research all the interviewees were in agreement that control over an internal R&D function was crucial, and this included the divisional companies. This implied strategic control, that even divisional companies broadly enjoyed, though one supplemented by access to resources within the parent. Furthermore, where the parent company provided finance to divisional companies it was not to correct for imperfections in the external capital market, but rather, by way of hidden subsidies, it seemed to be investing in disregard of capital investment principles. Operational control at divisional companies tended to extend as far as the production systems which, for historical reasons, remained within the company structure. For example, Bentley manufactured its old V8 engine internally but W12 engine production was more dependent upon VW. At the same time, all the divisional companies had some claim to being full-function integrated organizations.

Although the divisional companies could not claim the complete full-function structure of independent companies such as Honda, they had enough to be able to claim similar governance structures. Honda in particular had a centralized governance that was more U-Form than M-Form, even if overseas functions were often permitted a small degree of autonomy. Divisional companies such as Jaguar and Bentley had far more control over their strategy than any overseas Honda function; so although they were dependent upon their parent companies for support there was an argument for describing their governance structures as U-Form based on their strategic control. This is not to say that the parent groups are not involved in divisional strategy, and Jaguar seems to receive a significant amount of strategic involvement from Ford, but this never seems to be consistent. Hedlund's (1994) N-Form of governance structure might be applicable, but there is little sense amongst the contextual companies of the heterarchical structure that the theory proposes. Instead, the empirical evidence appears to suggest that the U-Form has been adapted to the larger companies, perhaps facilitated by some kind of matrix that created temporary networks of management according to the project, whether it was a new car program or an overseas production plant. The dominant governance structure was therefore found to be the U-Form, either with international extensions for firms with overseas facilities, or with subsidiary U-Form governance structures for divisional companies.

5.5 Conclusion

The investigation of transaction costs has shown that there are two main dimensions to vertical integration: physical assets and human assets. Hitherto, physical asset specificity has been considered to be the main driver of vertical integration, internalization of these assets being a method of pre-empting *ex post* contractual opportunism. Although there is empirical support for this view it fails to recognize that human assets pervade all company activities, including control of the physical assets. Indeed, opportunism is essentially a human problem, not a technical one. This is because the decision to create a hold-up and appropriate *ex post* advantages is a human one, which then uses the physical assets to extract the benefit. Klein, in particular, has argued that vertical integration is a matter of internalizing labor contracts in order to introduce greater flexibility into the production system.

Assuming that companies using the same production technology are operating in the same economic environment, since they are therefore faced by the same transaction costs they will converge on a single form of vertical integration. This is the full-function model which determines the functions that should be internalized within the company boundary. For an automobile manufacturer this should include R&D, BIW, powertrain, and final assembly. It is an economic model of vertical integration and it does not denote a particular governance structure or geographic location of function. However, by juxtaposing theories of economies of scale with that of transaction costs it is possible to provide for each function prescribed levels of scale. This results in a model of size and vertical integration structure, but not governance structure.

Vertical integration brings with it control problems that different organizational governance structures are intended to address. According to theory, the unitary U-Form is effective for relatively small firms where the management can take a close operational and strategic role in the integrated functions. Although small firms are unable, by their nature, to exploit the available economies of scale, the simplicity of their structure brings this problem to clarity. The theory of the multidivisional M-Form isolates senior management so that it can specialize in strategic matters and leave operational decisions to the various divisions. The structure has advantages over transacting in the market because the relationship between headquarters and the divisions is transparent, thereby facilitating efficient allocation of resources and capital. This has particular utility when the industry is undergoing rapid development or a company is particularly diverse. The divisions themselves may be single functions or have a U-Form style structure with a number of functions, although without the overall strategic control. The M-Form can then bring benefits from economies of scope in the product range and management capability, although this can overshadow the pursuit of economies of scale, particularly in administrative functions. There are also doubts about whether the M-Form can stretch to an international structure.

The data gathered by this research was very supportive of the need for an integrated structure. This was even to the extent that there was more support for the U-Form than the M-Form structure. This is in line with the notion that the parameters of the industry are fairly stable and that the industry has reached maturity, a level of stability that Burton and Obel (1980) and Donaldson (2001) suggest is suited to the U-Form structure. Even large, international companies, such as Honda, maintained a strong focal center in the domestic market which showed a U-Form structure from which international extensions radiated. Although the foreign operations appeared to have operational control, this was strongly influenced from the center. For those firms which were divisional companies, such as Jaguar and Bentley, they too showed a U-Form integrated structure within a larger group, but this time at a much lower level of scale. This might have implied a holding company H-Form of governance, except that the parent company had occasional strategic and operational involvement in the divisional companies. At the same time, the parent company had its own automobile production facilities that somewhat accorded to the U-Form of organizational structure.

It should be cautioned, though, that the companies so far discussed have been close to, or exceeded, the prototypical format in both structure and scale. Even the divisional companies were able to exploit economies of scale by virtue of their membership of parental groups. The next chapter will explore the methods by which a company that has not reached the prototypical level, and so is uncompetitive within the automobile industry paradigm, can use devices to approximate to the paradigm. Thereafter, in Chapter 7, empirical evidence relating to such approximations will be introduced. In particular, the chapter will focus on how MG Rover had a full-function structure but lacked the scale to sustain itself. It will be suggested that the company could have used its human asset endowments as the basis for an organizational restructuring, which would then have allowed it to retain the transaction cost advantages of a vertically integrated, full-function organization while obtaining access to the necessary economies of scale. The reasons for the failure of MG Rover to achieve these benefits will also be examined in Chapter 7.

CHAPTER 6

Approximating to the automobile industry paradigm

The automobile industry paradigm described up to this point suggests strict conditions for the long-term sustainability of an automobile firm since it is constrained by parameters set by both economies of scale and vertical integration. So far, I have put forward a prototypical size and structure for an automobile firm operating closest to its optimum. This applies principles of economies of scale to indicate plant sizes which, in combination, provide a size for the firm. TCA is applied to indicate a full-function model of vertical integration. The full-function structure is effectively independent of firm size since it is related to the economic friction that exists between operations, thereby driving the make-or-buy decision.

In this chapter I will first examine the cost disadvantages of falling below the required levels of output and then demonstrate how these might be ameliorated internally with different production strategies. In Chapter 3 I showed that if the MEPS of the automobile company functions are brought into reasonable synchronization then this would result in the prototypical automobile manufacturer producing around 600,000 units a year at the MES_P. It has been found that beyond this optimum point diseconomies of scale emerge, partly due to rising R&D costs. However, firms producing at the MES_P are sensitive to changes in output that lead to large variations in costs. As a company grows in output by adding new plants, each designed to produce at the MEPS, this sensitivity is reduced so the firm is able to absorb changes in output without suffering the large variations in cost. Furthermore, large firms can diversify risk by offering extended product ranges. These firms are the industry leaders.

Those firms that fall below the MES output are deficient in structure, scale, or both. In this chapter I will investigate some of the strategies by which a suboptimal company might approximate to the cost advantages of the MES described by the automobile industry paradigm. This will include a deconstruction of the full-function model to show how it can be reassembled in a vertical joint venture that approximates to the advantages of the internalized prototypical model.

6.1 Introduction

Firms that fall below the output required for the MES_P suffer debilitating cost disadvantages, relative both to the prototypical model and to the industry leaders. Such firms will be deficient in at least one area: either the scale of output is too low or they do not operate with a full-function organizational structure. They then have to solve their predicament by approximating to the precepts of the paradigm to put them on a convergent trajectory with it. This is possible using internal or external solutions. Internal solutions could involve more intensive use of limited production facilities through flexible manufacturing of multiple models, which then reduces the company's exposure to changes in market demand by diversifying the risk. Alternatively, the company might extend the use of the facilities by perpetuating the output of products with unchanged specification for longer periods of time.

In another approach, the automobile firm might seek external solutions by contracting with other firms. Clearly, in the make-or-buy dichotomy, this can mean seeking resources in the open market. However, this is not an approximation of the paradigm but the antithesis of it. Since the full-function model demonstrates cost advantages in its integrated organizational structure, any firm that sources externally as an alternative to the model will suffer relative cost disadvantages. In order to approximate to the full-function model a firm must find a form of intercompany relationship that represents an alternative to the make-or-buy dichotomy.

Reflecting the two forces that structure an automobile company, there are two ways in which this external partnership can come about. The first is size, where the partners are uncompetitive in terms of quantity. A full-function manufacturer that is uncompetitive in this regard is unsustainable only by relative degree: the further behind the standard set by the industry leaders, the less sustainable it is. This can be corrected through a pooling of joint resources with another company, although some of the advantages may be offset by additional transaction costs. For consistency, I will refer to them as horizontal joint ventures as it involves the enhancement of existing capabilities through sharing with a partner. Naturally, it is a strategy that is also of incremental value to industry leading producers, further improving their relative competitive positions.

A second form of partnership is structural. Here, the company has some functions at the desired size, but it is not a full-function organization.

Such a firm is more than simply uncompetitive within the paradigm: it has no prospect of sustainability as long as it is lacking functions – these must be sourced externally. It is possible to do this contractually in the open market, but the full-function model demonstrates that this brings transactional cost disadvantages. To approximate to the paradigm there is the opportunity to develop a VJV, a complementary relationship where the two parties bring distinct capabilities to the alliance. An additional strategy may be necessary if scale is still lacking, but if the parties already enjoy competitive economies of scale in their existing functions, then the resultant VJV will be converging upon the prototypical model and so have access to most of the available economies of scale. VJVs have less value for firms that already comprise full-function structures because the strategy implies relinquishing one or more of the functions.

In this chapter I will show that for firms to come together in a VJV each must have a different structural basis in order to have arrived at their contrasting structures. In particular, the distinction between human assets and physical assets in the automobile industry paradigm implies a demarcation line within the full-function model that might be exploited for the VJV. Specifically, human assets prevail in R&D, while physical assets prevail in production. This distinction between asset types may be heightened by the possibilities raised by globalization. New locations offer a range of different factor endowments, as described by the Heckscher-Ohlin (HO) model, and elaborated by Dunning's eclectic paradigm; some of the objections to globalization will also be noted. A firm that is already operating at, or above, the MES_P will receive incremental benefits from the different factor endowments to enhance its position relative to its peers. However, for a firm that is at a suboptimal level of output a restructuring around different factor endowments may raise it to a more competitive position. Thus, not only does an IVJV have the potential to replicate the organizational structure, and even scale, of a prototypical automobile company, but it can also bring additional benefits due to specialization.

Throughout the discussion examples will be given of cases that illuminate the theoretical approach so far described. These will develop the potential for joint ventures as forms of organization beyond the make-or-buy dichotomy. Globalization will be examined as a potential source of cost benefits due to factor endowment differentiation, particularly with regard to the division between human and physical assets. This will be applied to the joint venture structure and then further refined to give rise to the IVJV. In Chapter 7 I will then examine the empirical research data based on the theoretical principles established in this chapter.

6.2 Subscale production

The technical advantages of scale production are well documented; Pratten (1971) attributed further benefits to companies that achieve very large volumes:

- R&D costs can be spread over a greater number of units thus encouraging greater research.
- Large firms can expand incrementally in plant sizes that each achieve the MEPS, depending on the process.
- A large firm can save costs on designing new plants due to learning effects from previous projects.
- Larger firms can negotiate discounts on larger procurement contracts of components by extending the benefits of scale into the supply industry.

Many of these points were covered in Chapter 3, although R&D costs were actually found to rise as firms grew in size.

Exploitation of economies of scale may be a fundamental aim of the industry but, as Pratten (1971) has shown, it is not a condition that can be arrived at immediately. Even with a given level of technology, learning effects will influence actual output, so firms joining the industry now are at a greater disadvantage than if they had begun at the start of the industry. The Ford company that exists today was actually Henry Ford's third attempt at manufacturing. The vast domestic US market had already proved fertile ground for Olds with the first mass produced automobile and even the famous moving production line was not new (Lewchuck 1987). Yet, though Ford was aiming to increase efficiency, at every turn it is not clear that the concept of an optimum output was understood. Ford advanced the technology and techniques for production flows to improve production synchronization with the leveling of production times, yet this strategy is unrelated to the concept of an optimum output that would exploit the economies of scale for individual processes (Tolliday 1998).

The Ford company attributed its success to its production methods but did not consistently quantify this advantage. The Rouge River plant was designed to be a highly vertically integrated production facility with its own blast furnaces and glass making facilities. Thus, along with the latest company labor practices, the new plant was intended to be the ultimate expression of Fordism. This formula was repeated for the Dagenham plant in the UK. However, it proved beyond the capabilities of the company to match the minimum cost outputs of the different facilities within the plants so that all production activities could be operating simultaneously at the

MEPS. The Rouge River plant never reached more than 50 percent of its capacity utilization (Williams *et al.* 1994), and where once 100,000 workers produced 1,200 cars a day there are now 3,000 devoted to producing 800 Mustangs a day (*Economist* 2002). Fiat copied the old Ford Highland plant when it built its production facility in Turin, but this simply imported all the original associated problems along with a multistory architecture unsuited to flow production.

In the short run, it is important that companies operate plants at minimum cost output and that the resultant cost structure is comparable with that of competitors. In the long run, companies are at liberty to choose larger plants, on condition that the actual output accords to that indicated by the industry LAC. According to Maxcy and Silberston (1959, 93) the main benefits of a move towards an optimum scale are yielded in the initial stages. At the lowest end, expanding from 1,000 units per annum to 50,000 units per annum, there is a yield of a 40 percent fall in production costs. A less substantial 15 percent is available in the move to 100,000 units per annum, and the advantage slows further in the next expansion to 200,000 units per annum with a 10 percent saving in production costs. To achieve another 5 percent saving, output would have to undergo a dramatic increase to one million units per annum.

Of course, the MES_P put forward in this book is in the middle of the highest range quoted by Maxcy and Silberston. Pratten (1971) also suggested that the largest benefits in economies of scale are gained in the 100,000 to 250,000 range for a firm producing one model, economies available above that range being those attributed by Pratten mainly to body pressings. Cusumano (1985, 216) was another who found that Nissan achieved its greatest savings with the first expansion in output, finding a 31 percent decrease by raising production from 23,000 in 1955 to 130,000 in 1960. However, as the company grew so did the model range and by 1999 the company had 24 global platforms, five of them taking 62 percent of output and only one totaling over 300,000 units a year, for a total output of around 2.5 million (*Automotive Design and Production* 2004). I have found that beyond production capacity of 600,000 units a year the cost advantages can be attributed to diversification of risk, of which a burgeoning product range is a crucial part of the strategy. However, Nissan's example shows that this too can lead to rising costs if it is not controlled.

If the major gains from pursuing economies of scale are found by smaller companies that are expanding, it is also possible to mitigate, to a certain extent, the scale requirements of the Budd BIW process by using stamping dies designed for economic production at low volumes. The short-lived revival of Jensen, the British sports car manufacturer, was predicated partly on utilizing soft resin tooling which reduced die costs from £40 million (US$58 million) to £1 million (US$1.4 million), suiting a planned production starting at 300 units a year (*Financial Times* 2001). This is only a small saving, since all the other associated costs, such as the presses, remain the same. If the presses are capable of being run at 2 million operations a year then there will be an output mismatch if low volume dies are fitted.

There was another option available to Jensen that might have ameliorated the effects of subscale manufacturing. The marketing solution would have been the cost recovery tactic where the final product is priced to cover the additional costs of suboptimal manufacturing (Wells and Nieuwenhuis 2001). With a published price of £42,650 (US$61,420) in 2001 the Jensen SV8 was considerably more expensive than the comparatively high volume Alfa Romeo sports car (the GTV V6 being priced at £25,240 (US$36,350)) but similarly priced to other low volume sports cars (e.g. the Noble M12 GTO, priced at £44,950 (US$64,730)) (*Autocar* 2001). Although Pratten (1971, 147) stated that "a higher price can be charged for a non-competing article," it is unlikely that there would be no competition whatsoever, even if it is limited to another sports car or a substitute product; the existence of any alternative would mean that costs still have a crucial role. In any case, price recovery serves only to support high costs of production, it does not address the underlying cost problem itself. This may be achieved, though, by approximating to the automobile paradigm using internal or external resources.

6.3 Internal approximations to the automobile paradigm

Automobile firms will commonly exhibit a full-function structure including the MEPS for particular plants and processes. With an MES_P offered at a relatively accessible capacity of 600,000 units a year, firms are able to expand capacity in units of the MEPS. However, if a company finds that its scale of output is short of the MES_P, it must find an approximation to it. As a full-function firm there are two internally generated solutions available: either to prolong the use of the production facilities over time, thus reaching for the ultimate output levels over the long term but not the rate of annual use; or to exploit economies of scope by using the facilities more intensively with flexible manufacturing techniques and a broader product range. Although neither tactic achieves quite the optimums prescribed by the theoretical economies of scale, they do at least represent a pragmatic approximation.

Prolonging the use of the production system

Prolonging model life cycles can approximate to achieving maximum panel press usage by attaining ultimate levels of output over time, for example the figure of 7 million pressings for the life of a die (see p. 52) could be achieved by keeping the same model in production for a number of years. The most famous example of this would be the VW Type 1 (Beetle) which can claim production from 1938 to 2003. It achieved an annual peak of 1.3 million units in 1971, well short of the 2 million annual optimum for pressing rates alone (as opposed to the BIW process as a whole), but it made up for this with a total production run of nearly 22 million that somewhat exceeded the prolonged life strategy. Other examples would be vehicles such as the Jaguar XJS, which remained in production largely unchanged for 21 years, even though total production only amounted to 112,000 (Buckley and Rees 2002).

Maxcy and Silberston (1959) cautioned that no matter how much the life of the dies is prolonged, this cannot correct for the disadvantage of running heavy press equipment at low rates. In addition, it does nothing for the operating rates of other functions, although powertrain production is usually boosted by the fact that powertrains have the potential to be shared with other models. Recently, though, it has become increasingly difficult to prolong engine production due to the progressive tightening of engine exhaust emission regulations. Prolonging model life cycles is therefore a decreasingly appropriate strategy.

Economies of scale in flexible production

Economies of scope elsewhere can have a direct impact on production and the exploitation of economies of scale, the enabling technology being flexible manufacturing systems. Flexible techniques have become a feature of all three of the manufacturing functions, since multiple products diversify the risk of output fluctuations. For example, a wide product range means that a high volume manufacturer, like Honda or Toyota, can use one production facility to make relatively low volume, niche products. In final assembly this may be relatively easy to achieve as long as the different vehicles can physically join the production line and the routine nature of vehicle assembly means that the same workers can work on different vehicles with little additional training. Powertrain production is also amenable to flexible production of multiple products, for example engines tend to be designed and made in "families" that offer a range of sizes. However, since these techniques depend on the manufacturer having a wide range of products on offer then it tends to be a strategy better suited to larger manufacturers, those operating at the MES_P or higher.

As discussed in Chapter 2, a multiproduct process seems to offer economies of scope. However, the use of common components has the result of raising the utilization of the production system and, as Clarke (1987) noted, this only serves to improve access to economies of scale if it employs production capacity that might have been used less efficiently elsewhere. Few manufacturers achieve more than 50 percent usage of their body presses in panel stamping (Maxcy 1981), so the ability to use a wide number of die sets sharing a limited number of presses by swapping between the sets will raise the capacity utilization rates of the presses. This assumes, of course, that the changeover time is short enough. This does not quantitatively change the economies of scale that are available in panel stamping but it does make them more accessible by keeping the presses running for longer, even if the die utilization is unaffected. This is further improved by minimizing the time necessary to change the dies, often possible in just a few minutes (Cusumano 1985, 285). Such producers also benefit from having a flatter total cost (TC) slope due to higher mechanization and lower variable costs, reducing the break-even point and leading to a rapid increase in profits, though also rapid increases in losses if the strategy fails (Rhys 1972, 276).

Although flexible production is an opportunity to access economies of scale, it also introduces costs of its own. Even if presses are run more frequently by alternating dies for different models, the time taken to switch dies implies additional costs and does not address the cost of the dies themselves. Nieuwenhuis and Wells (2002) put a total BIW investment cost at between £370 million (US$560 million) and £515 million (US$770 million) for one model, including £20–£65 million (US$30–US$93 million) for the dies (see p. 51), so with the simple addition of a new set of dies for a different model, using otherwise the same facilities, the BIW process would be in the range of £390 million (US$590 million) to £580 million (US$870 million) for the two unrelated models. Since some of the unseen panels may be shared, the total costs can be reduced, but mixed production can never exploit all the available economies of scale, if only because both sets of dies cannot be used simultaneously. Nevertheless, this does at least represent a significant improvement over running plants below the MEPS.

One company, Fumia Design, has even proposed the use of symmetrical outer panels, for example matching front and rear doors, thereby reducing the number of dies necessary per model (Fumia Design 2006). GM found further advantage in its C-Flex system for welding the panels together.

The company uses welding robots that can be simply reprogrammed for different models instead of having to be replaced, resulting in cost savings of $100 million (£67 million) per vehicle launch, and a plant area reduction of 100,000 sq. ft (*AutoTech Daily* 2002).

Mehrabi *et al.* (2000) suggest a novel form of flexible manufacturing, the reconfigurable manufacturing system (RMS). These authors see the RMS as a quantum advance over flexible manufacturing because it includes the possibility of adjusting production capacity to fit market demand. This implies access to economies of scale for a wide range of outputs rather than achieving a single optimal output with a mix of models. This would be of interest to all manufacturers, but particularly a prototypical manufacturer seeking to reduce the risk of fluctuating output, and also those below the MES_P seeking to reduce the optimum level of output.

The RMS is achieved by modularization of the production process in terms of both hardware and software. As consumer demand changes, so one system can be reconfigured into another by the expedient of replacing, deleting, or modifying selected modules. However, for such an approach to be effective, the operators would need a resource pool that they could draw on, or deposit redundant modules into, without incurring additional costs. Essentially, an RMS reduces production capacity by removing machines, but the authors fail to explain how this can be done without leaving those machines idle, and therefore incurring costs. Furthermore, it is difficult to understand how the remaining machines could be employed in a reduced production process unless they were inherently flexible. Under the Mehrabi *et al.* (2000) system this would not be possible since modularization is contrary to generalism: it only applies to specialist machines.

From the description by Williams *et al.* (1994) of the TPS, developed by Taiichi Ohno in the 1950s, it is possible to visualize the underlying logic of an RMS in the creation of U-shaped subassembly cells, essentially spurs feeding into the main production line. However, Williams *et al.* (1994) pointed out that this led to duplication of equipment. Furthermore, the purpose of the cells was not to bring flexibility to production but to achieve uninterrupted production flows, bringing the rate of subassembly work up to the same speed as the main line. This would then inhibit, or eliminate, the buffer stocks that build up between machines that operate at different output levels, as described by Ho *et al.* (1983). The same situation that is being controlled by buffer stocks can also cause blockages when the buffer is not large enough (Gershwin 1987) or stoppages if the buffer becomes exhausted. TPS subassembly cells constituted an organizational innovation but they had no effect on the design of the main line, which continued with linear production flows dependent upon automatic transfer machines (ATM).

Since final assembly operations are characterized by a sequence of activities rather than a specific technology they have attracted many innovations to improve the production flow. Womack *et al.* (1990) have argued that Toyota has shown superior use of the production system by developing lean, or fragile, production. The most well known development is that of JIT, whereby inventories at the assembler are kept at the lowest possible levels, necessitating deliveries from suppliers at time-critical intervals (Krafcik 1988). This involves close relationships with suppliers, the primary ones being termed Tier 1. At the Honda plant in Ohio, for example, 92 percent of Tier 1 suppliers provide JIT and design contributions, with Bellamar producing seats in parallel with the vehicle for which they are intended and supplying them direct to the production line (Mair 1994).

The JIT system depends on a self-regulating pull-through production system governed by "kanban" cards and "andon" lights, with the machines and subassemblies organized to minimize labor. However, Williams *et al.* (1994) took the same view of JIT as they did of TPS in general: indeed, they termed it the "romance of Toyota." Instead, the authors put the spotlight on Shotaro Kamiya and his Toyota Motor Sales (TMS) division, created using selling techniques learnt in the United States. To keep the production system running it was necessary to maintain the stability of demand, and this, according to Williams *et al.* (1994), was the crucial contribution to production made by TMS to TPS, rather than the innovations within the plant itself. By 1975, Toyota had overtaken Nissan and it continues to hold this position, taking 40 percent of the Japanese car market in 2005 by consolidated sales (*Automotive News* 2006a). Toyota has promoted the advantages of the "*genbashugi*," a management technique that emphasizes the importance of making operational decisions at the production site. However, according to Williams *et al.*, only a successful marketing strategy feeds through the sales volumes required for production to operate at high levels of efficiency. In Chapter 3 it was shown that a major source of cost was the fluctuation in output from that which was planned.

The TPS, therefore, represents a refinement of the existing production method: the production line is still a linear sequence, there is large scale use of ATMs, and the *kanban* system is similar to the system of human chasers used by Ford at the Highland plant. Williams *et al.* calculated that final assembly by Toyota makes up only 14 percent of the total cost of production anyway, the rest being borne by suppliers using unrelated methods of production. Furthermore, in its home location, Toyota delegates much of its vehicle assembly to affiliated companies in which it holds equity shares ranging from 100 percent (wholly owned) down to 23.51 percent.

Williams *et al.* asserted that Taiichi Ohno understood this primacy of maintaining a smooth production flow by production leveling, flexible manufacturing also helping to achieve this by assembling multiple vehicle models. Refinements such as this have led to the progressive improvements in output reported by Rhys (1972, 1999), but not the quantum change that would signal a paradigm revolution in car production. The Nissan plant at Sunderland is recognized as the most efficient in Europe, yet it is neither quantitatively nor qualitatively unique (Times 100 2004) and so the MEPS is largely unaffected. This indicates that the continuing development of production systems has increased the ability to exploit the economies of scale, but not necessarily redefined the overall MES_P for an automobile company.

If a company finds that internal solutions of prolonging or intensifying production are not available then it can try external solutions. In the make-or-buy dichotomy the simple option is to make contractually regulated transactions. The problem with this is that it introduces the very transaction costs that the full-function organizational structure avoids and thereby permits those costs to persist in a structural form. Since the company is attempting to get as close as possible to the full-function model expressed within the automobile industry paradigm then an external solution should comprise more than simply transacting for resources in the market. An alternative way by which it can approximate to the automobile industry paradigm is through alliances with other automobile manufacturers. The nature of such potential alliances will be discussed in the next section.

6.4 External approximations to the automobile paradigm

External relationships that are contracted for in the market are part of the make-or-buy decision process and do not represent a novel form of relationship that might approximate to the automobile paradigm. This describes the supply industry but it can also extend even to vehicle assembly under contractual conditions. For example, the contract assembler Magna Steyr assembles eight separate models for three different automobile companies at its plant in Graz, Austria. These models are variants supplementing production of mainstream models made by the automobile manufacturers themselves. There is a degree of asset specificity since the production facilities are specialized, but the bilateral relationship is balanced by the intellectual property rights being retained by the client automobile companies. Despite an output of 250,000 units a year the plant is fed by six body plants and two paint plants when one each would normally suffice.

The firm is not, therefore, benefiting from access to economies of scale and cannot do so as long as its clients insist on the separation of processes by product range (*Automotive News* 2006d). The clients are not exploiting economies of scale through Magna Steyr either, though presumably there are cost advantages over producing at their own plants. Furthermore, the relationship does not represent a novel form of organizational structure since the plant supplements client-owned assembly plants elsewhere. In essence, Magna Steyr is little different to the tier 1 subassembly suppliers, albeit resulting in fully assembled variants of mainstream models. The company is thereby supporting the automobile industry paradigm but it is not operating within it.

When two companies form an alliance they are side-stepping the dichotomy of the make-or-buy decision by entering into a conjoined activity or joint venture. The structure of the alliance depends on the extent that the activity can be divided between the partners. If the activity is not divisible then the partners will pool their resources into one jointly held facility. This is a horizontal, or lateral, alliance where both parties are able to expand their common capabilities in unison. In this situation, both companies are comprised of integrated functions but are seeking to share one or more of them in order to increase output, such as when entering a new market. For example, the NUMMI assembly plant in California, started in 1984 (JAMA 2006), operated as a joint venture between Toyota and GM, allowed the Japanese company to share in the scale benefits of the facility as it developed manufacturing in the US. In the preceding section I demonstrated how an internal problem of scale could be overcome with an internally generated solution, such as flexible production. An alliance or joint venture, however, provides an external solution in the form of a partner.

However, if the activity is divisible then it opens the possibility of the partners taking complementary positions, each company retaining its unique capability, in order to form a vertical alliance. In such a situation, the partners are in search of a capability that they had not previously internalized but which had perhaps been supplied to them externally through transactions. The vertical partnership might be intrafunctional, each partner contributing subsidiary skills within a function, and so each company gains access to a new facet of the range of related skills. The two companies are individually full-function but are seeking augmentation within one or more functions. Alternatively, when neither company has a full-function structure the partners might specialize in complete, self-contained functions within a multiproduct production system. When a vertical alliance or joint venture of this nature occurs it is interfunctional. The partnership offers the possibility of the two companies sharing in a combined full-function structure that grants access to complementary functions that are entirely novel to each of them respectively.

Hybrid organizational structure

Williamson (1991) considered the possibility of an interim form of company organization which neither made its transactions on the market nor regulated the transactions through an internalized hierarchy. This intermediate mode he termed the hybrid, which is a structural form made up of an eclectic mix of the two polar opposites, market and hierarchy. However, it is not at all clear how this renders the hybrid structure a distinct form of organizational structure in its own right. The author gave, as an example, the case of the contract between the Nevada Power Company and its fuel supplier, the Northwest Trading Company. This is described as a neoclassical contract because it anticipated *ex post* opportunities, in effect it recognized that opportunism could occur and so enshrine mechanisms for dealing with it when it did so. At such a point the "excuse doctrine" could be invoked, relieving the parties from strict legal enforcement of the contract terms. In this way it escaped the rigidity of a classical contract that fixed an *ex ante* agreement which was then enforced by law.

There is not, though, a clear demarcation between classical and neoclassical contracts. If no contract can be complete, and all are open to some *ex post* exploitation, then contracts of varying corruptibility should be compared as if on a continuum, not as distinct options. Artz and Brush (2000, 338) found in their research that, instead of a choice between discrete forms of organizational structures, there was a continuous range. This being so, the so-called neo-classical contracts are then simply contracts that are more complete than others because they cover a greater number of contingencies. If items in the contract carry no legal status then they are not binding and only the risk to reputation remains. This may offer an opportunity for arbitration, yet this would exist anyway, whether it was specified in the contract or not. In economic terms the pertinent issue is whether there is a legal contract between the parties, or whether the transaction has been internalized into an integrated firm where "hierarchy is its own court of appeal" (Williamson 1991, 274).

When Williamson put forward franchise relationships between companies as being hybrid forms he was describing how the two parties have greater autonomy than in a hierarchy, though this then entails greater use of rules than in a market in order to protect the interests of the parties. The relationship between a franchisor and a franchisee appears to be unique because of the complexity of what is being supplied. This includes product branding, quality, pricing, and even employment policies. A franchise agreement is therefore one that regulates a multitude of transactions between the parties by bundling them into one contract. Yet the essence of the contractual relationship remains the same; it is only that in this situation the contract has been written in recognition of a complex situation. It is therefore unlikely that the hybrid structure is one based on a new kind of contract, all contracts being written to regulate specific relationships anyway.

Cooperative organizational structure

Many commentators have suggested that firms can work together in a manner that is less dependent upon the mechanisms of the market and more dependent upon a spirit of cooperation and a sense of shared destiny. This concept of working together to mutual benefit may represent an alternative to the choice between internalization or market-based transactions and it has been approached from various standpoints.

If a company's planned output is located in that part of the LAC curve that exhibits increasing returns to scale then, rather than produce by itself, it can elect to form a joint venture in order to produce in a larger plant one that approaches more closely the MEPS output level. Kogut (1991) pointed out that joint ventures are useful in mature industries where the LAC is already well known. Excess capacity can be eliminated at the planning stage by combining production capacity with a partner, so that together economies of scale can be exploited.

Cleeve (1997) looked at Japanese firms entering the UK. Due to the indivisibility of certain production technology the incoming firm might find itself having to invest in more capacity than its intended output required. Naturally this would put the firm in a position where increasing returns to scale existed, and hence higher unit costs than at the MEPS, a situation which could be enhanced by partnering production with another firm. Hennart and Reddy (1997) also found that Japanese firms tended towards joint ventures when economies of scale were larger than intended output, particularly if the incumbent firms were already producing near the optimum.

Kakabadse and Kakabadse (2000, 114) have looked at firm service activities which, being generic in nature, were amenable to joint ventures or consortia: "the key objective of such arrangements, therefore, is to achieve economies of scale through a beneficial local source delivery to its consortium members." Weijian (1991) noted that additional benefits came in other areas, such as corporate learning, which could have a significant impact on shifting the SAC curve as personnel become more adept at operating the existing plant. Learning between partners indicates that cooperation brings its own benefits, Wilding and Humphries (2006) putting forward the concept of cooperation as an alternative form of organizational structure that might minimize the need for contracts or vertical integration. This would then support the possibility of approximating to a full-function model in a manner that replicated internalization and did not involve transacting externally. They found that market-based negotiations brought increased costs when they became adversarial in nature. The authors advocated "C^3 behaviour," explained as cooperation using fewer suppliers over a longer term, coordination of knowledge exchange and collaboration in planning.

Like Williamson's hybrid form, C^3 seems to underestimate the role that contracts have in the relationship between two firms. Although the benefits of C^3 behavior are quite apparent, it is not clear how it can be put in place *ex ante* between new partners without some kind of enforcement. Instead, C^3 behaviour is the *ex post* benefit of aligning transactions with the appropriate organizational structure. In a previous paper, Humphries and Wilding (2004, 1119) acknowledged that this condition of empathy between organizations could not be arrived at immediately – trust had to be nurtured until the full benefits of the C^3 relationship could be realized: "organisations should attempt small, simple, co-operative projects that improve efficiency because these are perceived as being non-threatening: discussions about costs should be left until some maturity has been achieved."

This seems to concede that cooperation can be developed if the incremental steps are low risk, and thus the transaction costs are also low. However, this serves the purpose of maintaining the equanimity of the partnership, such that one partner cannot assume a significant advantage over the other. Additionally, there is no reason why each "simple" step should not be enforced by a "simple" contract, thus spreading the transaction costs of specifying the ultimate contract over the relationship development period. At the other extreme, trust can have a malevolent aspect, as when one partner has a disproportionate amount of power which it can be expected, or "trusted," to use to its own advantage. Blois (1972, 268) argued that when a supplier depends to a large extent on one customer then it is quasi-internalized because of the customer's bargaining power: "in some cases large customers are prepared to indicate that they regard their suppliers as being extensions of themselves."

The concept of regulating company relationships by a spirit of trust in place of either contracts or internalization has often been applied to Japanese companies. Dyer (1996) used the principles of TCA to look at the *keiretsu* form of alliance structure seen in Japan, particularly in application to the automotive industry. These are vertical buyer–supplier relationships, so the *keiretsu* form is described by the make-or-buy decision of standard TCA, though it results in an alliance form of organizational structure. I have found that around 38 percent of component costs in Japan could be sourced to the *keiretsu* alliance, as opposed to the United States, which was much more polarized with 10 percent attributable to partnerships and 48 percent produced internally – the remainder being sourced in the market.

One explanation put forward for this difference is that American companies can call on a legal system that is more closely adapted to the characteristics of market transactions, while Japanese companies are culturally oriented towards a trusting relationship. However, this places too little emphasis on the *keiretsu* structure. Gilson and Roe (1993) pointed out that a *keiretsu* is a form of organization that anchors a network of companies to a common core, often a bank, and each company is a part owner of the network through a system of cross-holdings. This means that suppliers and buyers in the network are also equity owners, creating a degree of integration that lowers transaction costs and permits knowledge transfers. In a subsequent publication, Dyer (1997, 548) reiterated the role of trust but acknowledged that "a stock ownership position held by an automaker acts as a financial hostage which encourages the supplier to make partner-specific investments." The asset specificity is therefore founded on a sense of joint ownership and acts as a self-enforcing influence on the relationship: "self-enforcing safeguards can control opportunism over an *indefinite time horizon.* Conversely, contracts control opportunism for only a *finite time horizon*" (ibid., italics in original).

Toyota is often cited as a company that has achieved minimal cost in manufacturing by pioneering lean production techniques. Yet this is an approach that is readily copied by rival companies. In fact, the most interesting feature of Toyota is its highly unusual structure in its home market. There, as of March, 2006, the company has 22 major plants covering a range of functions from forging of parts to final vehicle assembly. Within this total, 12 plants are concerned with final assembly and yet only four are internalized. The remaining eight assembly plants are subsidiaries and affiliates, one of them being wholly owned, five being majority owned, and two being minority owned (Toyota 2007b). All of these external operations depend on Toyota for all, or the vast bulk, of their sales, and as such Toyota has influence over them as if they were internalized. It can even direct them to make investments in new plants, such as the new Central Motor assembly plant being built in northern Japan at Toyota's behest

(*Japan Times* 2008). Clearly, this dependency suggests an indefinite, self-enforcing time horizon rather than a danger of opportunism. It is not, though, based on cooperation without enforcement.

Exploring this approach further, an alternative form of joint ownership might involve retaining specific assets within a jointly operated production system, particularly if the technology is readily divisible. Pires (1998) looked at the relationship between the vehicle manufacturer and suppliers at the VW truck factory in Resende, Brazil. This plant is based on a consortium approach where production is divided into distinct sub-assembly modules. Seven dedicated suppliers have sole responsibility for production within each technologically separable module and must guarantee delivery to the final assembly line. There were hopes that the model could be extended to automobile plants, but there are two crucial differences. Firstly, the techniques used in truck production lend themselves to subdivision; secondly, truck manufacturing is a relatively new activity for VW. Pires also reported that the company had found it more difficult to institute such a style of production at its new Brazilian engine plant in São Carlos. Despite this, the model does illustrate that asset specificity is one that can secure a bilateral agreement when the partners are codependents using distinct technologies.

Mariti and Smiley (1983) recognized that firms may share specific activities through horizontal cooperative agreements. This is not a make-or-buy decision because this is not a supplier–buyer relationship. The two parties have already internalized the activity and wish to share it in order to exploit further benefits. In particular, this allows them to access economic advantages: "through co-operative agreements, firms can take advantage of economies of scale in one or more of their production processes while remaining separate entities" (439).

Although the authors argue that this is not simply a market transaction since the agreement excludes other firms, and therefore changes the competitive landscape, this is also the purpose of contracts negotiated in the market. Even when a cooperative horizontal transaction is non-monetary, often involving an exchange of knowledge or pooling of expertise in a joint venture, it remains a transaction even if it does not include physical movements of cash. For example, in 1998 Ford and PSA (Peugeot-Citroën) agreed to work together on the new range of small diesel engines, the DLD family (*Diesel Progress* 1999). Both companies had existing diesel engine design capabilities but in combining their experience they were able to realize economies of scope. This is often the reason for engaging external engineering consultancies, since it widens the breadth of skills that are available for the project and the Ford–PSA relationship might be seen to be in the same vein.

Link and scale alliances

If the cooperative modes of interfirm relationship so far described fail to offer an alternative to the make-or-buy dichotomy, Dussauge *et al.* (2004) have elaborated on the technological basis of the dichotomy and provided grounds for a new type of organizational structure. The authors categorize collaborative relationships into scale and link alliances. Scale alliances combine similar resources in order to seek economies of scale in a horizontal joint venture. Link alliances, by contrast, are oriented towards knowledge acquisition and combine complementary skills to allow the partners to access skills they have not developed for themselves. This opens access to economies of scope, but it has ramifications for the longevity of the partnership: Dussauge *et al.* have found that link alliances tend to be more volatile and result in one partner acquiring the other in order to secure long-term access to the overall knowledge base. This is because, the authors argue, link alliances are based on the *ex ante* uncertainty of potential knowledge which, when gained, accrues to each partner separately and gives rise to the possibility of asymmetric *ex post* opportunities. Naturally, this is the kind of behavior Williamson warned about in transaction costs and reveals the central role played by human assets in link alliances. Alternatively, scale alliances involve joint learning and continuing efficiency considerations, which would be lost if the partnership were dissolved.

In practice the distinction might not be so clear cut. For example, one partner may have the necessary skills but be lacking scale, and so from that firm's perspective the alliance is about finding scale through a pooling of resources. On the other side of the relationship the partner may be seeking economies of scope by access to its opposite number's wider knowledge base, and so will view the relationship as a link alliance. This can be the driving force behind the complementary partnership of a VJV. When the alliance is a link–scale compound form, it may be that the manufacturing strong partner will prefer to break off the alliance once it has found the knowledge it was seeking, since it will not need to internalize its partner's production capability. This will then leave the knowledge-strong partner marooned without access to scale production. In sum, the link aspect deals with a specific problem which is then solved through the alliance, while the scale aspect deals with a problem that is endemic to the system and which is only solved by its continuing operation. Although link alliances may be, by their nature, short term, I will show later that a link–scale alliance in a VJV can be sustainable depending upon the structure. The next subsection will demonstrate the crucial role that human assets play in business relationships, with implications for link alliances.

Human assets in link alliances

Klein (1988) revisited the relationship between Fisher Body and GM which had become a centerpiece in the transaction cost argument (see p. 113). The author tackles the idea that contractual imperfections lead to opportunism and hold-ups that can only be resolved by vertically integrating the two parties into one firm. This is particularly relevant to link alliances where the asymmetric development of the relationship can lead to one partner holding an *ex post* advantage over the other.

The original contract between Fisher Body and GM had secured the relationship by guaranteeing Fisher Body exclusive dealings with GM and thereby reducing Fisher Body's asset specificity risk. GM was in turn protected from price hold-ups by a price protection clause in the contract. Joskow (1985) found similar contracts concerning coal supplies for power generation, though there would have been far less likelihood of technical innovation changing the nature of the industry. The equanimity of the Fisher Body–GM agreement was disturbed by the exponential increase in demand for closed bodies which Fisher Body was in a position to exploit for its own gain.

Klein asserted that the long-term contract actually caused the problem by constricting the relationship during a time of change. The essential feature of the relationship, though, was not the physical assets owned by Fisher Body but the human assets, which by their nature cannot be owned. The human assets can only be controlled by employment contracts and it was at this level that vertical integration occurred, according to Klein (1988, 207): "it is in this sense of owning a firm's set of interdependent labour contracts and the firm-specific knowledge embodied in the organization's team of employees that an owner of a firm can own the firm's human capital."

Vertical integration therefore has little impact on the number of contracts, but since it involves absorption of the organizational capital of the other company the nature of the relationship is changed. It was no longer necessary for GM to prespecify how physical assets should be managed, such as production locations and output; and although the requirement for their use remained without the binding force of legal constraint there was far greater flexibility. Similarly, a link alliance constrains the main driver of innovation – knowledge – which is held by the human assets. If a firm is seeking specific knowledge then it will quit the link alliance when it has acquired it sufficiently. Alternatively, if it has a continuing need to develop the link alliance more than its partner then it will internalize the activity. Although the Fisher Body–GM relationship might appear to be a scale alliance enforced by a contract, in fact it had the character of a link alliance providing GM with access to steel body technology. Since GM's future depended on being able to direct the human assets involved it had the greater need to internalize the operation.

In cases where one partner is seeking specific knowledge then opportunism manifests itself in the form of early termination of the relationship once the knowledge has been gained. For example, Hill *et al.* (1990, 119) emphasized the risk that is attached to entering foreign markets by the mechanism of licensed production. The authors cite the example of RCA having its color TV technology expropriated by Japanese licensees. The picture is more complex when the relationship is a link–scale compound. When Toyota commenced production in the US it did so with the NUMMI joint venture plant with GM. Although this provided both companies with scale it also permitted them to learn about each other's manufacturing techniques. As predicted by Dussauge *et al.* (2004), the link alliance aspect was not extended, as Toyota went on to establish wholly owned production sites, though the scale aspect was strong enough to keep the joint venture plant in continuing operation.

A compound link–scale alliance that seemed to hold long-term potential in the UK was that between Rover Group and Honda. In 1979 Honda was still a young company, having begun automobile manufacturing in 1963 and exports to the US in 1970. There had been an early attempt at a European production plant in 1963, manufacturing mopeds in Belgium, but this had proven to be more of a challenge than the company could manage: "the Belgian experience was one of hardship in every aspect of work, including production, sales, development and management" (Honda 1999, 137).

The venture exposed the lack of market knowledge and poor adaptation of the product to European tastes. The experience seems to have impressed on Honda its lack of experience in the region. Indeed, its comprehensive lack of economies of scope in management, R&D, and product range was quite conclusive. The subsequent introduction of the N600 small car into the US market appears to have been rather tentative, being initially released into Hawaii. The contiguous US was entered with greater commitment with the introduction of the Civic in 1973 and then the Accord in 1975 (Honda 2004).

After the apparent debacle in Belgium, the automobile market in Europe was also approached with caution. Honda lacked economies of scope in manufacturing knowledge and marketing to which it could gain access via a link alliance. At the same time, the financial damage caused by the Belgium production site made a scale alliance in a shared production facility equally attractive until sufficient demand justified a wholly owned operation. For Rover Group production, capacity was not an issue, but it was in need of vehicle designs. Although the company probably had the strategic resources in R&D at the time to design its own vehicles, the vicissitudes of the previous few years had conspired to leave it with a tactical need for a new model to maintain its production and market presence in the short term. For Honda it was therefore both a link and a scale relationship; for Rover Group it was purely a link alliance.

The link–scale theory of Dussauge *et al.* (2004) would have anticipated that the link–alliance aspect would last only as long as the need for knowledge persisted, while the scale aspect would endure. As will be discussed in greater detail in Chapter 7, Honda soon gained economies of scope from its experience of European manufacturing and markets, and the creation of its own assembly operation by 1989 gave it greater access to economies of scale in addition to sharing production with Rover Group (JAMA 2006), just as had Toyota and GM. This was not reciprocated by Rover Group, which revealed a chronic lack of economies of scope in R&D, being capable of developing its own designs for some vehicles, particularly at Land Rover, but barely able to transfer these skills to its mainstream models.

As Dussauge *et al.* (2004) assert, a link alliance becomes destabilized once just one party has gained the knowledge it was seeking. It seems that the Honda–Rover alliance was anchored by the physical asset specificity of the scale aspect in joint manufacturing, and that this secured the continuity of the relationship in order to maintain the human assets of the link aspect. As Chapter 7 will show, the alliance was ended by the acquisition of Rover Group by BMW in 1994, which broke the scale aspect of the Honda–Rover relationship, and the Japanese company was able to extricate itself from the link aspect. Even so, from that point Rover Group, and its successor MG Rover, retained the Honda-sourced technology until the company's demise in 2005. However, without access to subsequent Honda developments via the link alliance the product technology gradually fell behind the competition.

The complexity of compound link–scale alliances suggests that firms will work together for a variety of reasons, many of which they may not share in common. Firms are not simply motivated by scale or knowledge: there are other factors involved, such as differences in the production environment, particularly when the companies operate in spatially diverse regions with inherent factor endowments. Globalization has increased access to these factor endowments and this suggests that the IVJV may encompass link–scale considerations while offering the possibility of a new form of organizational structure outside of the hierarchical–transactional dichotomy. The possibilities raised by globalization will be discussed in the next section.

6.5 Opportunities raised by globalization

Globalization of trade has brought economic benefits by allowing countries to exploit their comparative advantages and make better use of their factor endowments. The mechanism by which this is achieved is specialization which allows nations to focus on their strengths and trade the surplus in return for those goods in which their trading partners specialize. Although this is a simplified view of international trade – for example, trade will also involve exchanges of goods which are unavailable to the importing country – in this section I will show how international trade has accelerated over the past few decades. However, this trend has also brought with it a great deal of criticism.

The development of international trade based on the principle of free markets in turn offers new opportunities to companies. Instead of being part of an international exchange system, companies can become enmeshed in the economies of other nations through FDI. For this to occur, the investing firm needs to hold some advantage that the domestic firms do not possess. At the same time, the foreign location must present an opportunity to the foreign firm that exceeds the risks involved in operating in an alien environment.

Automobile firms have been early explorers of international locations but their heuristic evolution has highlighted the complexity of engaging in FDI. Companies that have attempted to internationalize find that they must account for the benefits existing in the foreign location as well as the advantages they can bring to it. Since automobile firms comprise a number of production systems, a single mode of market entry for the firm is not applicable. International factor endowment differentiation can mean that the firm needs to approach the new market in a selective fashion, with perhaps only certain processes being amenable to FDI.

Automobile firms may also find that globalization presents opportunities that outweigh disadvantages that it had been suffering in its home market. It might allow a firm to access factor endowments that may favor certain functions over others but which were not readily available in the home base. In this way, globalization opens the possibility of geographically dividing the full-function integrated structure according to the availability of factors of production. For a company that is operating beyond the minimum output (MES_P) prescribed by the automobile paradigm, such a strategy would be of incremental benefit unless the company restructured the existing full-function structure around the new possibility. For such a firm, at or beyond the MES_P, FDI is used to enhance advantages it already possesses. The possibilities for a firm operating well short of the optimum prescribed by the paradigm, though, are much greater because of the opportunity to restructure around a new location. Furthermore, because the opportunities are applied only to discrete processes, depending on whether they are capital or labor intensive, it creates the greater potential for vertical relationships rather than horizontal. In this section the nature of globalization and the opportunities it offers for automobile companies will be discussed, showing that the IVJV is the organizational structure that best approximates to the automobile paradigm.

Globalization of trade

One of the most powerful arguments for the gains to be made from the international movement of goods is the HO model. This uses differences in national factor endowments to explain how countries are biased towards either capital or labor intensive production, assuming that the underlying technology is commonly accessible. Although this should lead to specialization by countries in production to suit their factor endowments, this is inhibited by diminishing returns. For a country endowed with labor the progressive attraction of that factor to production will lead to the factor price of labor being bid higher. Conversely, the factor price of labor will fall in a country endowed with capital, just as the factor price of capital is bid higher. The two national trends will converge until factor prices have equalized and the conditions for international specialization have been removed. Schott (2003, 686) argues that this implies that international industrial development has an end point for each technical process: "this single cone version of the model has all countries of the world producing all goods, so that both Japan and the Philippines, for example, are assumed to produce identical electronics and apparel goods using the same techniques."

Based on empirical evidence, Schott (2003) takes the view that industry develops in a dynamic manner but always originating from its factor endowments. Convergence is therefore a tendency, not an absolute. Schott's "multiple-cone equilibrium" would have the Philippines tending towards production of labor intensive clothing and Japan tending towards production of capital intensive high-technology goods. This is even as factor price convergence pulls in the opposite direction with a tendency towards equilibrium in labor and capital. Although countries, or regions, rich in labor will attract capital to a degree, as new industrial developments emerge countries will revisit their persisting factor endowments.

Deepening the HO model in this way brings greater insight to the subtle distinction between the long established practice of internationalization and the recent rise of globalization. International trade has a long history and Peter Dicken (2003) describes how this has encompassed many commodities, such as spices, but adds that production processes used to be organized within national barriers. Trade took place as a result of the factor endowments described by the HO model with a flow of raw materials from the periphery to the core and a return flow of manufactured goods. It is by the mechanism of international trade that companies can internationalize, geographically spreading operations to take advantage of the different factor endowments.

Berger (2000) points out that there has existed a common understanding since the 1990s that the international economy has changed such that it is now embodied by a single market in goods, services, capital, and labor. As predicted by the HO model, the dichotomy caused by the factor endowment differential has been replaced by a blending of the two sides such that the periphery no longer describes a location, the site of raw materials also hosting a manufacturing base. Although internationalization worked through the mechanism of international trade, globalization removes the distinction between locations. As Sturgeon (2000, 3) states: "the underlying hyporeport is that the more that national economic systems come to resemble one another, the fewer barriers will exist to the flow of resources to their most efficient use, and the further the world economy will become integrated, or globalized."

Yet Sturgeon and Florida (2000) anticipate Dicken (2003) in finding that the ultimate state of globalization is, in reality, a continuing process, and that it has yet to supplant internationalization. Furthermore, Sturgeon (2000) points out that the degree to which industries can exploit globalization is uneven and the author finds evidence of globalization in capital markets, regionalization in supranational clusters of industry, and internationalization in national trade policies. Dicken (2003) also distinguishes between internationalizing processes, which are characterized by quantitative increases in economic activity between nations, and qualitative globalizing processes, which include the integration of economic activities across nations. On this basis, while global markets are measured quantitatively the globalization of production is essentially qualitative (Sturgeon 2000).

Applying this to Schott (2003), quantitative increases in international trade will not only bring national industries to a degree of convergence, but the increase in opportunities will encourage further innovations that will again differentiate between relative factor endowments. Schott enriches this view of factor endowments by including factors beyond capital and labor to take in factor efficiency, demand, political policies, and so on.

Yet this complexity is not always well understood by the companies that seek to exploit the advantages, nor the governments that seek to control it. Concern has been voiced on the changes that globalization is bringing and how companies might use it to challenge existing political authority.

Critics of globalization

There is little doubt that the globalization of international trade has brought many economic opportunities. Hill (2003) has considered the influence of capital flows, as manifested by FDI, quoting the United Nations' figures of $60 billion (£26 billion) in 1980, $210 billion (£118 billion) in 1990, and $732 billion (£442 billion) by 1998. Hill infers that this shows the rise of international production systems and the spread by organizations into the markets of their foreign competitors. The caveat to this is that it involves only a select number of countries, with ten developed countries receiving up to 70 percent of the world's FDI in 1999. In the same year, Africa received a meager 1 percent of the world's FDI (Hill 2003, 6). Rugman and Hodgetts (2001) also show that while the majority of international trade is between the triad of the North American Free Trade Agreement (NAFTA), the European Union (EU), and Asia, an even greater proportion of the trade is within the individual elements of the triad.

Cowling and Sugden (1998) suggested that the inequality of trade patterns show how trade is not so much free, more that corporations are free to act across international borders. In this context, freedom means the ability to implement strategic decisions. Since these firms are controlled only by those who have a direct vested interest in it, the strategic decisions that drive international trade are made by the senior management of the corporations. Cowling and Sugden (1998, 347) concluded that "freedom in free trade is the freedom of an elite to manage international trade in their interests, despite the objections of others."

Cowling and Tomlinson (2005) state that the concentration of power in a few transnational corporations puts developing nations at a disadvantage. One such objection came from the Malaysian Prime Minister, Dr Mahathir Bin Mohamed, who laid the blame for the East Asian crisis of 1997–98 on the huge new international corporations: "their strategy is simple. Become so big that no one else can compete. The small must either allow themselves to be swallowed or suffer failure very quickly" (World Economic Forum 2002).

Some commentators believe that the perceived iniquities engendered by globalization will lead to retaliation. Polanyi (1963) argued against the commoditization of three elements of production: labor, land, and money. Essentially, the author sees these elements as being fictitious commodities: land, because it is not in production, and labor and money because they are not produced for ultimate sale. Indeed, they are fundamentally social activities, and so society rebels against their commercialization in a self-regulating market (Silver 2003). Furthermore, taking labor out of the market will bring about a transformation as great as the original establishment of the competitive labor market (Polanyi 1963). Munck points out that a worldwide expansion of capital will also lead to an expansion in the proletariat, increasing the importance of labor. Even if the workforce has not been globalized yet, "there is the tendency towards the creation of a global labour market" (Munck 2002, 11).

Countering this are social forces that would retain the differences in factor endowments. Trade unions such as Britain's Transport and General Workers Union (TGWU) are sensitive to the risk of jobs being moved overseas as a result of the opportunities presented by international recruitment (Benady/TGWU 2003). Cowling and Tomlinson (2005) note the existence of international institutions, such as the World Trade Organization (WTO), and the rise of consumer groups that seek to force a "corporate responsibility" program on corporations, but nevertheless conclude that the power of multinational enterprises (MNEs) to conduct their own strategies has merely been mitigated, not banished. Although Bartlett and Ghoshal (1993) suggested that the M-Form of corporate structure was unsuited to coping with resource allocation judgments for globally dispersed operations, the example of Ford seems to show that control from the strategic headquarters can be highly effective. The company's strategic management compared information gathered at individual plants as the basis of threats to the local employees for moving production elsewhere (Beynon 1984). The company was also able to use the information to put pressure on political authorities, one example being the 1977 engine plant in Bridgend, South Wales. Of the £180 million (US$314 million) required £115 million (US$201 million) came from the British government.

According to Barnet and Müller (1974, 41) the strategic management of a global company, as seen in the M-Form structure, is able to coordinate all the factors of production, whether mobile or immobile: "it is this ability of World Managers to weigh all these factors and to co-ordinate decisions on pricing, financial flows, marketing, tax minimization, research and development, and political intelligence on a global level that gives the world corporation its peculiar advantage and extraordinary power."

Hymer (1972) claimed that this represented a unification of world capital and labor which reduced the independence of nation states. This view is based on Hymer's theory concerning the manner in which MNEs can use the advantages they have developed in their own markets to exploit advantages in new markets, despite the fact that they are operating in an alien environment. This will be discussed in the next subsection.

Foreign direct investment (FDI)

Although the comparative advantage theory of international trade can trace its lineage back to David Ricardo in the eighteenth century, the manner in which firms can use it as an enabler for their own international structuring has only recently been articulated. Until Hymer proposed a new approach to FDI in 1960 the accepted view of such foreign commitments was that they were a form of portfolio investment (Rayome and Baker 1995). Portfolio investment committed funds which exploited differences in interest rates but without securing operational control (Hymer 1960). Hymer took the view that this did not sufficiently explain FDI, since funds for the foreign operation tended to be raised locally and, in any case, the resultant distribution of it did not follow the availability of low interest rate opportunities but industrial ones. Instead, Hymer concluded that firms would wish to have some control over the foreign operation in order to reduce competition and secure the full financial benefits of their capabilities. The funds that were raised locally to fund the investment were only benefiting from interest rate discrepancy by coincidence; the main purpose was to obtain sufficient funds to achieve at least minimum control of the project. Indeed, by exploiting a low interest rate environment for FDI, the flow of capital would be the reverse of that predicted by the theory of international capital movements (ibid.).

Hymer's theory arose from the imperfections in the international market caused by the lack of integration between national markets. This meant that firms could leverage the advantages they enjoyed in their home market, perhaps because they had developed there earlier, and apply them in new foreign markets where companies had yet to evolve to the same state or did not have access to the same resources. The internationalizing firm would thereby find additional advantage in the foreign market over its home location (Yamin 1991). However, this advantage could only be exploited by drawing on resources in both countries and the advantage would have to overcome the disadvantages of operating a business in an unfamiliar environment (Kindleberger 1984).

In situations where an existing local firm is, or has the potential to be, fully competitive with the internationalizing firm then Hymer contended that FDI would eliminate competition, perhaps through merger, acquisition, or some other form of collusion. This occurs when the firms are in an oligopolistic situation such that a coordinated relationship would bring increased rewards. Hymer's theory was further refined by Kindleberger in order to articulate the nature of the market imperfections, culminating in the Hymer–Kindleberger theory of FDI (Rayome and Baker 1995). These imperfections included:

- Departures from perfect competition in goods, for example product differentiation and pricing strategies.
- Departures from perfect competition in factor markets, for example patent laws, differential access to capital, and management skills.
- Economies of scale.
- Political intervention.

However, the Hymer–Kindleberger theory has been criticized for focusing on the endowments of the investing organization and for not including costs that might apply to such factors as the planning of the investment or the financial burden of the acquisition. The theory was therefore not predictive of the precise form of resultant corporate structure or how it came into being (Rayome and Baker 1995). This was because Hymer appeared to be unaware of Coase's theory of the role of transaction costs in structuring firms, a criticism that is not shared by Horaguchi and Toyne (1990). They point out that Hymer was not simply describing FDI in terms of the proactive exercise of market power, but also in terms of its being reactive in attempting to minimize costs through internalization. Hymer noted that this influences the governance structure of the multinational firm. In its local operations it needs to adapt to local conditions using decentralized decision making, while in its international operations it needs centralized decision making in order to coordinate the flow of information: "they must therefore develop an organizational structure to balance the need to coordinate and integrate operations with the need to adapt to a patchwork quilt of languages, laws, and customs" (Hymer 1970, 445).

This seems to suggest that the M-Form of decentralized governance would suit domestic firms, while the U-Form would provide greater control over the disparate foreign functions. The empirical evidence that was discussed in Chapter 5 similarly shows that international companies tended to use a U-Form with extensions into overseas markets.

Caves (1996) took a new perspective on Hymer's theory when observing that FDI tended to occur in a select number of industries. He noted that as the national firm considers FDI it needs to confront the costs of the expansion. Transaction costs imply that the firm will exploit opportunities in the domestic market until such a point that the marginal returns to its expansion begin to decline. The empirical data reviewed by Caves indicates that firm size is the most reliable indicator of the move abroad, yet firm-specific factors will have enough impact to render the decision details largely unpredictable. Since the internationalizing company carries disadvantages of being alien to the new environment, it will seek to minimize the information costs by approaching the most familiar foreign markets first. At the same time, it has already gained an information advantage, perhaps concerning production technology, in its home market, which can be used to offset the other disadvantages of operating in an alien environment (Calvet 1981).

Buckley and Casson (1998b) considered the theory by Caves to be static, since it focuses on firm-specific competitive advantage, choice of location, and determination of firm boundaries. Although this is not to say that Caves is in error, the theory is incomplete by not taking into account market volatility, the role of uncertainty, and managerial capabilities. Flexibility is the key to the actions of the company within an international environment that is in a state of flux, so the accompanying theories of FDI need to reflect the dynamic developments.

Dynamic theories of FDI

Vernon (1966) took the dynamic view of global operations in his theory of product cycles. The cycle follows the product from its initial development and introduction to product standardization and maturity. The theory considers matters of economies of scale, the timing of innovation, and the distribution of knowledge. The argument is that entrepreneurs in a location will have the best knowledge on products to release and thus innovation in that product will be initiated in its home market: "all of these considerations tend to argue for a location in which communication between the market and the executives directly concerned with the new product is swift and easy, and in which a wide variety of potential types of input that might be needed by the production unit are easily come by" (ibid., 195).

As the product evolves during its lifespan it passes through three main stages (Jones and Wren 2006):

- Product development process: product specification and production process is uncertain; production is located close to intended market.
- Maturing product: an established product and production process; economies of scale take precedence over location.
- Standardized product: a fixed product and production process; low pricing strategy takes precedence.

As an example, Vernon shows that high labor costs in the US have meant a growth in such innovations as automatic washing machines and drip-dry fabrics.

During the first stage in a product's development in the market place there are strong locational factors, such as a need for close control over the evolving product and communication with the market demand. As the product gains in popularity it matures into a standardized product which is generally acceptable, price then becoming the differentiating feature. Consequently, economies of scale come to the fore and production location may eventually move to a low labor cost region. Thus, as the industry grows, supply shifts from home production for the home market to export for foreign markets and then imports back to the home market from foreign production sites. Should the popularity of the product decline then companies will begin to exit the industry while the remainder restructure. Calvet (1981) saw the shift in emphasis of location as being a response to the loss of competitiveness as the uniqueness of the product is diffused, international expansion being the attempt to capture the remaining rent from the product's development.

Antràs (2005) takes Vernon's point on communication and analyzes this from the transaction-cost perspective, communication being expressed as friction and therefore incurring transaction costs. Whilst these costs are high, during the period when the product is still poorly defined and market demand has not yet coalesced, R&D and production are internalized. Even when the product has become standardized and production has moved abroad, Antràs argues that the move into the new market creates uncertainty due to the lack of knowledge and that this is then alleviated by FDI which maintains the internal functions of the firm. Vernon (1979, 256) similarly took the view that firms would reduce risk that was derived from their lack of knowledge: "firms are acutely myopic; their managers tend to be stimulated by the needs and opportunities of the market closest at hand, the home market."

Vernon (1979) emphasized that R&D tends to reflect the resources and market characteristics in which it is located. Yet even when there are cost advantages to producing in a foreign location, the firm may still not have enough information to make the decision to restructure. This may require a "trigger," a point at which a threat becomes clearly quantifiable. This threat may have varying relevance depending on where the firm is on the product cycle. Almor *et al.* (2006) divide the firm's activities into three parts: R&D, marketing, and production. They see R&D as determining the firm's competitive advantage during the early introductory phase, supplanted by the importance of marketing during the growth phase and finally production during the product's maturity phase. The locational advantages remain constant throughout but become relevant to different firm activities as they come to prominence. Early R&D, as Vernon noted, has a strong bias towards the home market, but as the product output grows marketing takes an international perspective followed by the internationalization of production.

Almor *et al.* (2006) note that with three factors driving the international growth of the product, internationalization of the company cannot be measured by the location of production alone. Indeed, a significant problem for Vernon's product cycles is when it concerns a class of product, such as electronic devices or automobiles, which are conceptually standardized yet are being continually updated. Klepper (1996), instead, preferred to take the industry as the unit of analysis and referred to the industry life cycle where innovation in product and process is developed concurrently. In such a circumstance, technical R&D is likely to be retained in its original home location while market-oriented R&D would be distributed throughout its target areas in order to develop local variants. Production might drift towards low labor cost locations but continuous R&D innovation in processes would still require a substantial presence in the home market.

FDI is therefore not a simple matter of systematically moving production at predetermined points in the product's cycle, although this may indeed be one of the factors to be taken into consideration. Buckley and Casson (1981) investigated the factors that need to be accounted for when deciding on FDI and found that in addition to the costs of servicing the foreign market it was also important to evaluate the demand conditions and future growth potential. Porter (1986) gave four factors that lead a company to seek competitive advantage through global expansion. These are:

- Economies of scale.
- Coordination advantages.
- Proprietary learning curve.
- Comparative advantage of the location.

However, these do not necessarily apply equally to all globalization situations. The first brings competitive advantage if it enhances the ability of the company to approach or achieve economies of scale, though this would inhibit an international move if the new production site was a substitute for existing output. Having coordination advantages is relevant only if it is applicable in the new location to a greater degree than the incumbents.

The proprietary learning curve is gained in the home location, but this can be disturbed by an international move that introduces new elements. For example, in an assembly operation the position on the learning curve will have been achieved after experience has been accumulated by the workforce, and although much of the technical learning can be transferred, the workforce in the new location will, nevertheless, have its own operational learning curve to climb. Under these conditions, a firm would expect to use its established home base as a foundation for exploring international expansion incrementally. This suggests that a firm would retain its existing governance structure as far as possible, such as the U-Form or M-Form, depending on the complexity of the organization, rather than implementing a new governance structure that was flexible enough to accommodate regional differences. Yet as Vernon (1979) observed with "myopic" management, the status quo preferred by the current industry leaders may inhibit them from taking the risk of restructuring and so destabilizing their competitive positions. This would involve the firm progressing in stages, establishing or learning at each stage before going on to the next. The Uppsala Internationalization Model takes one such behavioral approach, the firm deepening its foreign involvement due to the management push to learn and the pull of market demand (Buckley 2002). It is made up of four stages. These can be summarized as follows (Barkema *et al.* 1996):

- Stage 1: no regular export.
- Stage 2: exports through agents.
- Stage 3: offshore sales subsidiaries.
- Stage 4: overseas production.

Clearly, Stage 1 shows no significant internationalization, even if no firm is entirely insulated from the greater commercial world. Stage 2 takes a product into the international market via agents but has little effect on the operation at home (Andersen 1993; Viitanen 2002). Stage 3 brings a commitment to a specific overseas market, possibly with some custom product design, and the company might establish a degree of physical presence through franchising, licensing, joint ventures, and strategic alliances (Hill *et al.* 1990). The progressive internationalization culminates in Stage 4 where the firm is operating in the foreign market as an incumbent. At this point, the firm might claim to be globalized if the operations are integrated with those in the home base. Conversely, if the existing governance structure is retained, foreign operations might also constitute a simple extension of the domestic operation.

It should be emphasized that stage models describe the company ratcheting-up its exposure to foreign conditions, and this implies that it is going through different levels of commitment, though it would be facile to state that globalization is reached incrementally. Any corporate activity is incremental if the opportunities are fertile; to state that globalizing companies expand by stages is simply to describe the growth of success. The only types of firm that can avoid any kind of incrementalism and instantaneously globalize are those of the "born global" kind. Bell *et al.* (2001) have described such firms as small, entrepreneurial organizations that can draw on a sophisticated knowledge-based infrastructure, but this hardly describes the large-scale, technically mature automobile company described by the automobile industry paradigm. Aw and Batra (1998), in a study of Taiwanese manufacturers, found that small and medium firms used geographical diversity in place of the product diversity enjoyed by large firms. This indicates that FDI is not a strategy open only to those firms with a strong, home base but also those that are seeking to counter disadvantages at home or whose opportunities are global by nature.

Johanson and Vahlne (1990) conceded that there are exceptional circumstances when stages can be omitted because the learning is inconsequential or has been gained elsewhere. The three exceptions cover instances when company resources are large enough to render the risks of intermediate steps relatively containable: situations where the market is homogeneous and occasions when the market is similar enough to other markets for experience there to be relied upon. Each of these exceptions seems to describe large automobile manufacturers that can expand in units of the MEPS (i.e. constructing new plants that have all the size benefits of the MEPS) within a reasonably globalized market. Since the stage model theory focuses on the role of the company one might also add the opportunity to learn from other companies or first movers.

Johanson and Vahlne (1990) defended the stage model theory because learning effects account for a reverse flow of information which then influences the progress of internationalization towards globalization. This two-way process is part of the integration that characterizes the degree of integration necessary for globalization. Indeed, it may be a problem for any theory of internationalization that the nature of globalization is progressing and reducing the alien nature of foreign locations. This does not mean, however, that there is no differentiation between home and foreign locations, but it does mean that it has become more complex and subtle than Hymer originally envisaged. The eclectic theory, or paradigm, has arisen to draw together the different theories of FDI by applying them to different perspectives. This is discussed in the next subsection.

The eclectic paradigm

Although company learning must influence the progressive internationalization of a firm, the stage model theories do not specify what knowledge is necessary in order to embark on the next step. Not only are the theories not normative but, in allowing exceptions, they also fail to be completely descriptive. They therefore connect poorly with the HO model since they do not emphasize the complex differentiation of factor endowments.

Cheng and Kwan (2000) studied FDI in China and found that local market size, infrastructure conditions, wage costs, and political policies were all positive factors for inviting foreign investment into the country. Furthermore, FDI seemed to be self-reinforcing in that once the process had started it was easier to encourage more, suggesting that the FDI companies had their own internal factors for making the investment. This level of complexity cannot be contained within the foregoing theories of FDI but instead is better served by the eclectic theory which seeks to reconcile the different approaches (Dunning 1979). The eclectic paradigm is more able to encompass such complexity since Dunning (1979, 1981) formulated it to include factor endowments in three forms, known as the OLI parameters (ownership, location, internalization). These encompass the constituent approaches as follows (Dunning 1979):

1. Ownership advantages: industrial organization and ownership theories.
2. Locational advantages: theories of international trade.
3. Internalization advantages: theories of vertical integration.

This takes a broad view of factor endowments and includes various kinds of assets. Although the parameters are customarily referred to in terms of OLI, Dunning (1979) stated that for FDI to take place they are conditions which should be satisfied in a sequence better described by ownership-internalization-location:

1. The firm has possession of net ownership advantages, in the form of intangible assets, when compared to other firms in the same market.
2. It is then more beneficial for the firm to exploit these advantages internally rather than pass them on, by sale or lease, to an external source.
3. The preceding two advantages are better utilized in conjunction with a local factor input in the foreign market (e.g. natural resources).

Dunning (1980) summarized the OLI elements as the ability to call on resources that the competition cannot: the extent that the company's resources are better retained within its structure, rather than being obtained from an external source, and the profitability of exploiting these resources in conjunction with the indigenous resources of a foreign location. Dunning (1979) warned that the theory does not indicate which firms will engage in FDI or where it will take place, although Dunning (2000) identifies four main types of firm which can then be linked to other theoretical approaches:

1. Market seeking: Uppsala Stage 2.
2. Resource seeking: HO model factor endowments.
3. Efficiency seeking: HO model specialization (Schott 2003).
4. Strategic asset seeking: enhancement of ownership advantage (Erdener and Shapiro 2005; Bailey *et al.* 1994).

It is notable that the theory acknowledges that advantages are not evenly distributed, opening the possibility for more dynamic approaches to FDI, such as Schott's (2003) theory of international trade. The eclectic paradigm shows production processes can be separated into constituent parts and transferred globally in order to exploit differentials in factor endowments. As Schott demonstrates, this does not conclude the process of globalization since new developments will change the relevance of factor endowments, even within Dunning's OLI categories. This has already been observed within Vernon's product cycle, but it is tempered by the TCA approach which would inhibit the infinite division and dispersal of production processes, instead restricting this to technically separable processes.

This needs to be borne in mind since Dunning's approach, by the fact that it is eclectic, offers a myriad of possibilities. Williams (1997) emphasized that ownership deals with intangible assets, such as the access to parent facilities at a cost below the market rate in the new setting or the benefits of existing economies of scale. Dunning (1979) pointed out that a country might improve its stock of intangible assets by promoting a higher than average expenditure on R&D. This would have the consequence of raising its revealed comparative advantage. Ownership is also a control issue, comparable to Porter's (1986) coordination factor, introducing some sense of human asset specificity. Locational advantages are more closely related to the factor endowments described by the HO model, though Williams (1997) explained that these include, amongst others, trade barriers and tax regimes. For example, UK government policies have influenced choice of location, from intervention by way of the Industrial Development Certificate of the 1960s (Adeney 1988), to inducement by way of financial incentives later in the 1980s and beyond (Garrahan and

Table 6.1 Plant site selection criteria

Plant	Start of production	Reasons for site
Ford, Dagenham	1931	Logistical communications, labor pool
Nissan, Washington	1986	Existing site, local labor pool, government incentive
Honda, Swindon	1992	Existing site, logistical communications, labor pool

Sources: Dagenham, Washington (Georgano 2000); Swindon (Mair 1994).

Stewart 1991). Dunning (1979) explained that this means locational advantages are dependent on the fit between the characteristics of the location and the nature of the investing firm. Dunning (2000, 178) seems to anticipate Schott (2003) in noting a shift from the attraction of natural resources, or factor endowments of the basic HO model, towards "a distinctive and non-imitatible set of location-bound created assets."

Even these can change. For example, Ford of Canada was established to supply the nations of the British Commonwealth from inside the tariff regime (Wilkins 1979). Subsequently, Anastakis (2004) has observed the shift in ownership of Ford of Canada from a semi-autonomous operation to an internal division of the Ford Motor Company once the trade barriers between the US and Canada had been dismantled. Locational advantages can also include the existence of local industry clusters offering economies of concentration that the incoming firm can exploit using its existing ownership advantages (Buckley and Ghauri 2004).

Site specificity takes a narrower view of plant location advantages, including local rather than national characteristics. Table 6.1 contains a short summary of three automobile plants in the UK and the reasons given for the selection of site. The common features of this list are the existence of good transport communications, a large labor pool, and the availability of an existing industrial site. Nissan and Honda constructed new factories on the sites of old airports but with excellent road communications. Ford sought out port access and built on a greenfield site, more accurately described as a marsh, paying the price by having to build the foundations on 22,000 concrete piles and exceeding the budget by £2 million (US$9 million) (Collins and Stratton 1993).

The internalization element of OLI is directly related to TCA and the avoidance of market imperfections. This is pertinent when transacting across national borders where alien legal structures may raise the risk of opportunism and thus transactional costs. The make-or-buy decision facing the firm is solved by a range of solutions, from hierarchical, wholly-owned

FDI to market-sourced licensing and franchise agreements (Dunning 2000). The investing company would particularly need to protect its knowledge base, suggesting a tendency towards FDI (Erdener and Shapiro 2005). Bailey *et al.* (1994) suggested that this may be part of monopoly capitalism, emphasizing a strategic defense against oligopolistic rivals.

Strategic FDI

The eclectic approach provides an overview of a company's FDI activity within the context of the industry, but firms will also act in their own strategic interests. For example, although Hymer had rejected the portfolio approach to foreign extension as inadequate, Miller and Reuer (1998) found that diversification of exchange rate risk was a driving factor behind FDI since it provided significant hedging advantages over exporting. Cowling and Tomlinson (2005) mention the ability of multinationals to exercise strategic control over production, employment, investment, and advertising as a route to higher profits. For example, VW consolidated its two underperforming North American plants on its operations in Mexico so it could operate that one plant more efficiently (UNCTAD 2002). The attraction of low labor cost may also be weaker than previously thought. Erdener and Shapiro (2005) found that low labor rates are not crucial to the FDI decision and instead the motivation is towards the establishment of strategic assets, locating plants in order to more effectively service the local market rather than exploit cost differentials.

To be fair to Hymer (1960), the less well known FDI motivator in his theory was the desire to eliminate the competition. Buckley (1990) suggested that MNEs can use their international coverage to raise entry barriers to the industry, cross-subsidizing those markets that are most vulnerable to new competition. It is entirely reasonable that a company should have control over its functions, or else it could hardly lay claim to being a company at all. Yet if the purpose is for higher profits then it might be valid to suggest that strategic control is used to change the emphasis of the economic decisions rather than to take their place. This being the case, strategic decision making only has value to the degree that it brings benefits that exceed those lost through implementing a Pareto inefficient plant.

Should another firm find greater cost advantages within the OLI theory, perhaps by retaining a low labor cost location, then it could find itself at an advantage to the firm that had elected to invest in a higher labor cost location for a strategic purpose. In other words, if the strategic risk is miscalculated then the strategic decision will be rendered a disadvantage.

Buckley (1990) pointed out that in a static approach to FDI, for which the OLI theory has been criticized, organizational structures are likely to converge to an equilibrium, so it can be further argued that FDI based on strategic decision making is likely to have a more variable element to it. Buckley argued that internalization brings long-term advantages because the benefits are relative to the market, but competitive advantage is more transitory since it is relative to other firms. Strategic decision making might therefore be concerned with exploiting temporary competitive advantage. Since risk aversion depends on the decision makers within the firm it can be suggested that strategic decision making is a characteristic of individual firms and not a part of the underlying structure of the industry.

Strategy may also take precedence over labor costs as a technique for dealing with labor organization. Cowling and Sugden (1998, 76) discussed the strategic "divide and rule" concept whereby a company will eschew the cost advantages of maintaining a single plant and instead extend into foreign plants in order to prevent labor from unifying itself in opposition to the company. Whether this actually happens depends on the risk of labor being unified in opposition to the company and the degree to which the company is averse to this happening. There may even be an element of Williamson's concept of "atmosphere." Yet if this decision is a balance of strategic advantage and economic cost then it can only be made with knowledge of the cost considerations contained within the OLI theory. It is not an alternative decision-making process since any strategic decision implies a prior evaluation of the economic benefits of the alternative. Indeed, Table 6.1 shows that even if there were a strategic reason for the firms concerned extending into the UK as a production location, at the same time those sites had an economic foundation.

If a firm does engage in FDI for strategic reasons then it is entirely reasonable to expect a strategic response from rivals, a possibility that the OLI model of FDI would not account for since it is not concerned with strategy. The response may result in a defensive merger between local rivals or an aggressive extension into the FDI firm's home market (Jones and Wren 2006). This seems to argue that firms may make a strategic decision to invest abroad based on reasons that are in defiance of the advantages set out by the OLI theory. They may even engage in FDI to match the strategies of rival companies, Kuemmerle (1999) observing industry trends in the progressive internationalization of R&D. This opens up a multitude of possible strategies, such as strategic alliances that access economies of scale, partner specific expertise, or allowing a low risk entry into a new market. Buckley and Casson (1998b) added cultural factors, trust, and psychic distance to the more quantifiable factors of location and internalization seen within the OLI theory. Yet even with the additional sophistication of their proposed model, the authors conceded that it assigned a passive role to national governments which may have their own, strategic, reasons for encouraging FDI. On this level alone FDI can become highly complex, with governments setting defined limits to permitted FDI while the investing companies may perceive heightened risk in some countries over others.

Obtaining information is therefore crucial to reducing risk, but it is not unique to the strategic theories of FDI. Johanson and Vahlne (1990) criticized the eclectic model for assuming that managers have access to perfect information. It is the impossibility of achieving this that leads firms to be biased towards ownership advantages, about which it has accumulated knowledge from its own operations. Strategic FDI may therefore be a method for dealing with information deficiency. In order to preserve the advantages a firm has in an alien environment it can enter in stages, not in the systematic approach prescribed by stage models, but to progressively learn about the OLI factors.

Jones and Wren (2006) argue that the area of greatest strategic importance is that of knowledge, with particular reference to R&D. Borensztein *et al.* (1998) found that FDI was effective in the transfer of technology and led to greater economic growth, demonstrating why national governments tend to be enthusiastic supporters of FDI. A caveat to this, though, is that the benefits were only realized if the host country had sufficient technological resources of its own to integrate with the incoming advances. Where this is not the case the FDI is left isolated from its immediate industrial environment and the nation obtains fewer benefits.

This suggests a crucial role for international alliances and Dunning (1995) conceded that the OLI paradigm needs adjustment in order to explain how joint ventures can open an avenue to additional knowledge by forming a mechanism for extending the firm's ownership advantages beyond its formal boundaries. Alliances permit access to new information, but learning is a dynamic process, interactive with the environment, and alliances will only stabilize if they are structured to account for the destabilizing influence of new knowledge. Since knowledge is contained within human assets this represents the keystone of an alliance. Kuemmerle (1999) found that FDI in R&D sites was on the rise, but that the purpose was to extend existing R&D from the home base. If the foreign R&D operation was intended to augment existing knowledge then it tended to be tied to institutions such as universities, whereas if it was intended to exploit existing knowledge then it tended to be tied to a production facility.

Beamish and Makino (1998) expanded the typology of JV structures to reveal the complexity of international relationships beyond the customary view of a JV between a foreign firm and a domestic incumbent. The study found that these "traditional" joint ventures had the potential for superior performance but had a high failure rate. For Dussauge *et al.* (2004) this style of JV, dependent upon exchanging knowledge, might be classified as a link alliance and the increased possibility of opportunism would impose a short lifespan on the agreement. A stabilizing factor is suggested by HO and OLI theories when differentiated factor endowments provide the circumstances for specialization within the full-function structure. Promising though this approach is, it implies a radical restructuring, even deconstruction, that the strategic theories of FDI may not anticipate. Much of the reason for this is that firms are oriented to using strategies to exploit their existing capabilities, not to restructure around new ones. For example, I concluded from my research of survivor analysis that as firms expanded their product range and plant output the risk from fluctuating demand was diversified. For such firms additional overseas plants represented gains to the firm as a whole, rather than gains specific to the new plant. Conversely, smaller firms may use FDI to restructure as the antidote to a weakness in their position in their home market, perhaps because they are operating below the MES_P overall or the MEPS in particular processes.

To put FDI into an empirical context, in the next subsection I will investigate the progress Ford made in its early years as it learnt to tackle the complexity of evolving an international structure with different governance structures that reflected the opportunities.

Ford: a failure of global structure

As Burton and Obel (1980) stated, the U-Form of governance structure is suited to companies that are relatively small, a condition most companies are in as they first become established. Ford retained this structure as it grew from its Midwest base; and the production expansion that extended across the home US market, and then Europe from 1911 (Georgano 2000), involved assembly operations building up components shipped in from the main Detroit factories. As Tolliday (1998, 59) points out, the Highland Park and subsequent River Rouge plants were designed as predominantly component manufacturing plants supplying dispersed branch assembly operations. In 1921, of the total built-up vehicle output, only 9.5 percent originated from Highland Park, the rest coming out of the various assembly facilities. Ford opened 36 branch plants between 1912 and 1925 in order to assemble cars close to their local markets. Two of these, Chicago and Minneapolis, could assemble up to 200,000 a year, but many were at the 40,000 level and utilized the earlier stationary assembly methods.

As Ford developed its international expansion it would have been an opportunity to implement an M-Form structure so that local managers could be at liberty to procure components locally and adapt products to the local market. From 1912, local production of certain components was permitted for the UK plant at Trafford Park but at the same time Ford rigidly imposed its "American system" of work practices, such as fixed day rates of pay, and this brought the company into conflict with British work practices at that time (Lewchuck 1987). British managers attempted to ameliorate the American influence by adapting work practices to suit local conditions, and even after Ford in Detroit culled the local UK management in 1924 and installed a new team, it too found it had to change American methods to suit local conditions (Tolliday 1998). Yet Ford of US persisted in its authoritarian rule from the center. European sales were collapsing and the company fell from first to fourth in British rankings, behind Austin, Morris, and Singer, mainly due to Ford's failure to give the Model T an engine that suited the local tax regime, leading to the product being priced out of the market (Wood 1988). Most extraordinary of all, after World War I, the British arm was even compelled to produce only left-hand drive vehicles in common with the US (Tolliday 1998).

It would appear that a weakness of Ford's style of U-Form governance structure at this time was that it was inadequate at accommodating the dynamism of foreign markets. Vernon's product cycle would have forecast that the Model T would find a receptive market in the UK since it had set the standard for the industry elsewhere. In the event, neither the product nor the production system, both now icons of automotive history, were entirely applicable to an alien environment. As suggested earlier, the product cycle theory needs to allow for the standardization of the concept after which continuous adaptation and development takes place. In Ford's case the company missed its chance to adapt the Model T, but it did address this in 1932 when the company released the Model Y for the British market, a new product originally designed in Detroit. Its popularity induced the British arm to surreptitiously conduct its own limited R&D to adapt the product. This was comparable to the emerging M-Form structure then taking root at GM but at Ford it was instead met with a definitive restatement of control from the central management in the US. As leading Ford manager Charles Sorenson testily telegraphed to the chief engineer at the Ford Dagenham plant, A. R. Smith: "SMITH. ARE YOU AWARE THAT WE ARE CONTROLLING DESIGN OVER HERE. SORENSON" (quoted in ibid., 67).

With the U-Form structure held largely in place it was not until 1937 that Dagenham was allowed, despite continuing American resistance, to take on greater local powers and initiate a replacement for the Model Y of British origin. This became the Prefect, later joined by the Anglia variant, and Ford moved back up to third place in the UK rankings for car production (Wood 1988, 66). Up until then, the company had been conducting operations in the UK, quite overtly, as an integral part of the U-Form structure in place within the United States and with little reference to UK national differences. For example, local information such as conditions of tax regimes, labor practices, and market demand had been largely disregarded. The more conspicuous information had been transparent and the company recognized the need to minimize transport costs exactly as it had in the home market, and also to evade punitive local tariffs by assembling locally from supplied CKD kits. The unified, authoritarian approach of the U-Form had not been sufficiently developed to cope with the local complexities, only the more conspicuous matters of entering an alien environment, and it was only when Ford stretched to a more multidivisional M-Form that it was able to exploit opportunities specific to the UK market.

The move towards the M-Form allowed the British arm of the company to develop within the local environment while benefiting from central strategic support. In particular this might have been the economies of scope that can become available under the M-Form. Considering the state of development within the industry at the time this change in governance structure brought Ford into line with GM. However, my research has found that in the current industry the U-Form appears to have reasserted itself amongst manufacturers such as Honda and Toyota, albeit with sufficient adaptation to local conditions. This suggests that although the move to the M-Form by Ford brought benefits based on divisional empowerment the greater benefits would have come from adapting the U-Form to its international extensions. For example, the Model T began to fail on the UK market because it lacked the correct engine and steering configuration, not because it was inherently unsuitable. The Model Y seemed to correct the problem, being an American design adapted to local conditions, but thereafter Ford abandoned centralized control and gave the British division both operational and strategic control based on its own R&D capability. From this point the British operation seemed to take on a domestic U-Form of governance appropriate to its strategy of being a largely insular UK-based automobile manufacturer.

In the process of this evolution, no attempt was made to account for UK factor differences in the wider sense proposed by Schott (2003). It might be conceded that the economies of the US and the UK were closely aligned at the time so the differences would have been slight; nevertheless, there was apparently no indication that the Ford Motor Company was seeking to restructure around the opportunities raised in the British industrial environment. This is demonstrated by the lack of influence by the UK branch on the US operations as part of a dynamic interchange between the two branches. This does not mean that Ford was not aware of the principles behind the OLI theory, indeed the three factors seem to have been very much part of the FDI plan. The weakness was that the corporate governance structure was ineffective in fully evaluating the value of each factor, particularly that of location. Instead, the company seemed to act more in accordance with the strategic theories of FDI in taking a manufacturing position within the UK market which was not strong enough to offset the weakness in the underlying economic considerations. As was suggested in the preceding subsection, strategic FDI cannot be implemented in ignorance of economic factors.

If strategic approaches to FDI have a role it is when the industry is standardized and the market globalized, at which point the economic factors are embedded. For example, a company might establish a plant in a location to match a rival's FDI strategy. Since the production technology and degree of local R&D will by now be part of the industry's status quo the strategy will be concerned with competitive advantage (Buckley 1990) and the benefits will be incremental. The M-Form of governance structure might be sufficient for FDI of this level, but as Bartlett and Ghoshal (1993) suggest, it would be insufficient for accommodating the full range of opportunities offered by international environments, such as in the dynamic form espoused by Schott (2003). My research in the preceding chapter showed that the centralized style of the U-Form could be adapted by large firms to the demands of operating in overseas environments. An alternative approach, though, applicable to suboptimal firms below the MES_P, would be to restructure around the opportunities suggested by Schott. The next section will show how a firm can be restructured in recognition of these international opportunities.

6.6 Reassessing the full-function structure

The full-function model contains the four main functions for a prototypical car firm and, since the underlying cost factors are common throughout the industry, the full-function structure has become a feature of the industry (see Chapter 5). On the basis that integration means internalizing the responsibility for economies of scale then in principle the options for structuring are polarized: a function is either integrated or it is not. Alliances that take place at the subfunctional level, whether of the scale or link type, introduce additional transactional costs. Scale alliances seem to be more stable, but their benefits are static and the alliance lasts as long as the specific plant. Link alliances are apparently more volatile due to uncertainty in knowledge development leading to *ex post* opportunism. However, in a VJV, where the division of the link alliance is at a functional level, then the benefits of knowledge development accrue to the partner that has sole responsibility for that function. There is less possibility of opportunism and so the partnership remains bilateral. This in turn indicates that a VJV structured functionally is likely to be stable in the long term.

Although the theory of the VJV, and the IVJV version, has been laid out, it is by the nature of the alliance that empirical examples are difficult to source. For the VJV to be formed both partners would have to be in one of two conditions:

1. Have complementary functions that are deficient in scale which they are prepared to abandon in the interests of restructuring around a VJV.
2. Be deficient in the full-function structure to a complementary degree (i.e. the two organizational structures are able to dovetail in a VJV).

In the first condition a firm would need to have the courage of its convictions, and the trust in its partner, to restructure solely for the benefit of the VJV. In the second condition both potential partners would have to be operating without functions that competitors deemed crucial to long-term sustainability. They would therefore be in an unstable condition and dependent upon some kind of support not accessible to rivals (e.g. government funding, the consumption of internal funds for survival). The opportunity to form a VJV in the second condition would therefore only last as long as the support remained in place. In the following subsections I will examine examples of vertical relationships in the automobile industry. The first, between GM and A. O. Smith, is used to demonstrate the stability of a vertical relationship. The second, between Rootes Group and Iran Khodro, shows the possibilities in an IVJV.

Vertical joint venture at GM

General Motors was organized with a novel form of corporate governance structure, the so-called M-Form (Williams *et al.* 1994), a bureaucratic structure for dealing with the management of internalized functions.

Under William Durant, the company had been constructed from a collection of unconnected assemblers and suppliers for the same reason that Henry Ford had brought functions in-house, but the economies of concentration in GM's home base of Flint meant the company did not have to internalize processes to the same extent as Ford (Chandler 1990).

This was illustrated by the previously discussed occasion (see p. 149) of GM purchasing Fisher Body in 1926 (Coase 2000; Klein 2000), which may be contrasted with the following long-term contractual relationship with the functional supplier, A. O. Smith, a chassis supplier to GM in 1932. The two companies had been in a close, almost symbiotic, relationship for years, and Coase (2000) has found that even the production method was tailored to GM's needs, exhibiting the characteristics of de facto vertical integration. In a codependent, long-term relationship the interests of the companies are served best through the maintenance of the stability of the relationship rather than jostling for a position of dominance. Williamson argues that when asset specificity is high, "buyer and seller will make special efforts to design an exchange that has good continuity properties" (1981, 555).

The example of A. O. Smith shows that high specificity can enforce a bilateral relationship (Williamson 1998). Furthermore, because each company occupied a discrete function there was a reduced possibility of opportunism since any *ex post* advantages would accrue to the partner responsible for the function. Clearly this was not vertical integration as long as the two companies were under separate ownership, though the relationship was indeed a vertical one since A. O. Smith's responsibilities occupied an upstream position in the production process. Etemad *et al.* (2001) have found that when the supplier specializes enough a codependency arises and the authors' proposed "interdependence paradigm" states that this relationship can be stable over the long term.

The elegance of the solution to the internalization question by A. O. Smith and GM is in the manner in which operational functions could be internalized to the relationship. As independent companies they were not extending beyond their established competencies. They were also at liberty to take advantage of innovations when they occurred and compete within the industry for additional business. Later, when the all-steel unitary body was changing the industry, GM was free from the costs of restructuring a chassis production unit, this being an externality to GM that A. O. Smith had to deal with. At the same time, asset specificity was a mutual problem, A. O. Smith specializing in chassis production and GM having no such capability. This was the basis of the interdependency that enforced the quasi-internalization of chassis production. From this flowed

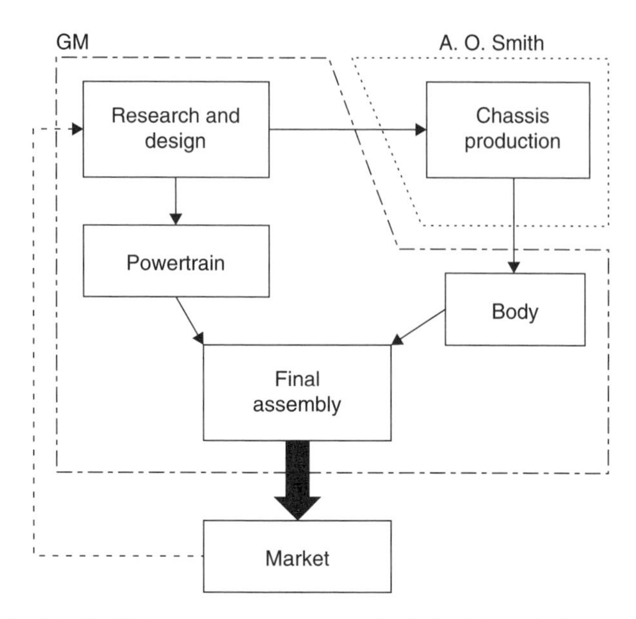

Figure 6.1 Vertical joint venture between A. O. Smith and GM

the benefits of internalization, such as a reduction in principal–agent costs of information searches, uncertainty, and contractual imperfections. With chassis output tailored to GM's assembly output, the production processes of the two manufacturers would have converged, thus the economies of scale for chassis production were accessible to GM.

The relationship between the two companies predates the Budd all-steel body paradigm, the chassis being manufactured using distinct technology: chassis production was a separate function from body production. A. O. Smith was therefore a functionary supplier, i.e. it had responsibility for the function of chassis production. Given the diverse range of activities that were held under GM's M-Form umbrella of governance it is interesting to note that A. O. Smith was not internalized. If economies of scope existed in the M-Form then it would be expected that they would be effectively exploited by such internalization. The fact that this did not take place demonstrates the value in the structure of the relationship in the form that it developed. This put A. O. Smith in a vertical position within the production process and in a close working position with GM, so although the relationship was not termed a joint venture it is reasonable to characterize it as such. Figure 6.1 illustrates the relationship between A. O. Smith and GM as a VJV. This has been adapted from the full-function model to show a pre-Budd era production system. It demonstrates A. O. Smith's critical position within the production process, the supplier's hold-up power being balanced by GM's status as sole buyer.

A VJV depends on the supplying partner having responsibility for the complete production process that defines the specific function. With the two being thus interdependent, the relationship is self-regulating and stable in the long term. A. O. Smith continued as a chassis supplier to GM until 1990, before finally moving out of the industry altogether in 1997 (*Assembly* 2002). However, this strategic relationship does not seem to fit any explicit form of governance structure. Although I have cast doubt on the efficacy of the M-Form of governance, particularly due to the difficulty in separating strategic management from operational management, in the case of GM and A. O. Smith this division appears to be more distinct. While A. O. Smith would have retained operational control, it is quite feasible to view strategic control as lying with GM, since it determined vehicle design and therefore chassis design as well as product demand. At the same time, the crucial feature of the M-Form, which is the transfer of objective data from operational units to headquarters for strategic analysis, would have been missing. From this one example a definitive judgment on the governance structure for a VJV cannot be made, but this may become clearer by investigating further aspects of the VJV relationship. Since it would appear that a VJV can be stable if the assets are divisible, it would be instructive to investigate how else an organizational structure, usually expected to be vertically integrated, might be divided. The following subsections will do this from the perspective of human assets and international opportunities.

Human assets and the full-function model

In Chapter 2 I compared the output of the three manufacturing functions (BIW, powertrain, and final assembly) with the R&D function, showing that diseconomies of scale set in early with R&D. Research has suggested that much of this is due to the rising staffing levels of larger departments. Clark *et al.* (1987, 766) stated that Japanese companies were able to counter this and find a relative advantage through project management: "in the best of the Japanese projects, a heavyweight project manager leads a multifunctional team, in which problem-solving cycles are overlapped and closely linked through intensive dialogue."

Although the presence of a "heavyweight" project manager implies a large team, it is noticeable that the multifunction structure indicates that the team is made up of smaller groups. As Cooper (1964) pointed out, a multigroup large team cannot entirely replicate the efficiency of small teams, and he noted that larger R&D departments have disproportionately higher spending levels. This results in decreasing returns to R&D investment and reveals the existence of diseconomies of scale. Within a full-function company the economies of scale captured elsewhere must compensate for the diseconomies in the design function or else the model would not be sustainable. Smaller companies, those with subscale production output, suffer the reverse effects. They might have small design functions that exploit economies of scale while the manufacturing functions are hampered by suboptimal production.

The difference between R&D and the manufacturing functions is the relative importance of human assets. Klein (2000) has shown that vertical integration is essentially concerned with securing human assets within a hierarchical company structure. Since R&D is concerned with the flow of new products to differentiate the firm, R&D must be crucial to the long-term prospects of the firm. It is therefore vital for the firm to control the R&D assets by internalization of this function and the human assets that comprise it. The volatility of link alliances shows the uncertainty involved in attempting to share this function with another firm.

The three manufacturing functions also make use of human assets but to a much lesser degree. As Klein (2000) has demonstrated, internalization of physical assets is concerned with gaining hierarchical control over the human assets that operate those physical assets. Thus, the physical assets of production are integrated for reasons of managerial control; the human assets of R&D are integrated for reasons of access to knowledge. In combination they form one discrete production system: the full-function model. Factor endowment differentiation offered by globalization indicates that advantage might be found in dividing the integrated structure according to the distinction between human and physical assets. The following subsection provides an example of an IVJV that demonstrates some of these attributes, and suggests a possible form of such a VJV that could have been beneficial to MG Rover.

International vertical joint venture: Rootes Group

In the 1960s, Iran was seeking to industrialize and had identified the car industry as a potential catalyst. The country was endowed with a large labor pool as well as natural resources, most particularly oil, which provided it with the funds to import capital. However, it had little existing industry and so was weak in engineering skills, the human assets of technology. It was therefore necessary for the company to import foreign vehicle designs for local production.

In 1967, the British company Rootes Group licensed the design and initially supplied CKD kits of the Hillman Hunter to Iran Khodro. In Iran, this model was known as the Paykan – Persian for Arrow – which had been the Rootes Group project name for the car. Rootes Group changed ownership over the next few years, in 1967 being acquired by Chrysler and then by Peugeot in 1979 when Hunter production in the UK was brought to a close. However, CKD kits were supplied until 1985 and then production continued with local component supplies. By 1993, the Iranian operation became 98 percent self-sufficient with output capacity of 120,000 units per year (Iran Khodro 2005).

Up to this point, the relationship between Rootes Group and Iran Khodro had been asymmetric in character, since there was a difference in the significance of the two parties. Under the terminology of the theory put forward by Dussauge *et al.* (2004), the relationship had aspects of both link and scale alliances. Access to the Iranian market provided useful additional output for Rootes Group, while for Iran Khodro it was principally a link alliance, providing access to vehicle design. As the theory of Dussauge *et al.* predicts, the scale aspect of the partnership was relatively long lasting with body panels being pressed in the UK, even beyond the end of the Hunter's production life. The link side of the relationship was very short lived, constituting simply the supply of the original vehicle design.

From Iran Khodro's standpoint the relationship could be characterized as a VJV, and this is illustrated by Figure 6.2. This shows Rootes Group supplying the R&D in the form of a finished product design, while Iran Khodro was engaged in the majority of the production. It is not, though, how the model would be depicted from the Rootes Group perspective, since the output for Iran was supplementary to all other functions and so would be described as a scale alliance. The mismatch in perspectives is partly explained by the HO model of relative factor endowments between the two countries and Schott's adaptation of it. Without vehicle designs Iran Khodro would have found it very difficult to conceive its own products, so accessing the engineering human assets at Rootes Group in the heart of the British car industry also suggested some locational advantages. For Rootes Group, though, the situation is complicated by the fact that production in Iran was simply an addition to the UK plants which were, at the time, fairly competitive within the domestic UK industry. Furthermore, as Schott (2003) suggests, factor endowment differentiation is complex and political factors had a great bearing on the partnership for the British operation. Government support in Iran of the nascent domestic industry would have made the abandonment of production in the UK by Rootes Group highly risky. Similarly, the eclectic paradigm shows that the locational

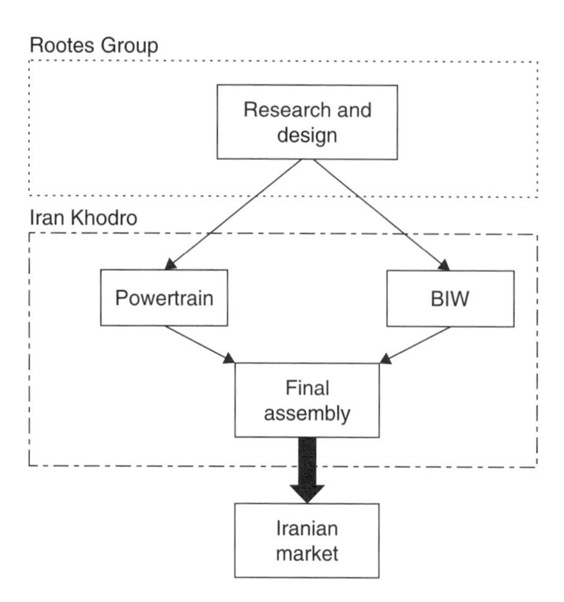

Figure 6.2 Vertical joint venture between Rootes Group and Iran Khodro: Iranian standpoint, 1993

advantages of production in Iran would have been outweighed by the risks of transferring production facilities to an uncertain political environment.

The role of the Rootes Group in the partnership was to solve the specific problem of Iran Khodro's need for a vehicle design. Once the design had been supplied, the R&D side of the Rootes Group had no more contributions to make, at least until the problem of a subsequent vehicle design arose. In the meantime, the Paykan was developed further by Iran Khodro using later Peugeot technology, the aging model being refreshed by building the old body style on a Peugeot 504 platform. The partnership was deepened when Iran Khodro designed its own vehicle, the Samand, on a Peugeot 405 platform (Iran Khodro 2005). Although the partnership with Rootes Group as a long-term scale alliance and a short-term link alliance had been in accordance with the theory of Dussauge *et al.*, given even more time in partnership with Peugeot the link aspect reasserted itself. Since Iran Khodro at least had the ability to develop the Paykan further, the need to replace it did not occur until sometime in the 1990s. It was then able to do this in a new link alliance with Peugeot, one unrelated to its link alliance with the Rootes Group. This indicates that in a link–scale combination alliance, the scale aspect secures the partnership in the long term and so allows the link aspect to continue from one project to the next, which Dussauge *et al.* (2004) show would not be the case for link-only alliances.

The scale alliance aspect continued because the problem of scale is endemic to the manufacturing system, while link alliances tend to provide solutions to specific problems. A scale alliance is therefore predicated on the nature of the physical assets, while link alliances are predicated on the knowledge contained within the human assets. In combination, link–scale alliances can exploit the divisibility of a firm into functions that favor either physical or human assets. Iran Khodro could readily act on this since it was lacking the R&D capability, but for the Rootes Group it would have been far riskier to abandon a production capability which was, at the time, still commercially viable. For this reason, the relationship can only be depicted as a VJV from the Iran Khodro point of view; for the Rootes Group it was a horizontal joint venture in production only. It cannot be said to have been a true VJV between the two partners because the Rootes Group never relinquished its production capability in favor of Iran Khodro.

However, as it was, it did provide some additional perspective on the divisibility of hitherto integrated company structures. The spatial division of human and physical assets, according to factor endowment differentials, can also be viewed as a type of link–scale alliance when it brings together R&D and production capabilities which are geographically separate. It might be considered a hypothetical question whether the M-Form of structure could have coped with internalizing Iran Khodro into Rootes Group, yet the reality is that the M-Form cannot be applied when national interests regulate against it. It is not simply that Rootes Group were denied by the Iranian government the opportunity to consider the question: the denial itself was part of that internalization calculation since it raised costs to an impossible level. It has been discussed previously that the M-Form is a bureaucratic device for accommodating diversity as long as it is limited in scope, for which it fails to account for the full range of opportunities that exist in international locations.

This then raises the question of the kind of governance structure that would be suitable for an IVJV. I have argued that the U-Form has the most suitable control mechanisms for an international enterprise, but it is not generally possible to have a centralized decision-making process within a JV. Indeed, an IVJV may require a particular governance structure of its own to manage the strategic structure of the combined operation. For further investigation, I propose that for a case study company, MG Rover, a link–scale combination alliance could have been structured along functional lines in an IVJV, each partner maintaining control of discrete functions such that any *ex post* advantages would accrue to that partner. The empirical evidence for the potential of such an IVJV will be discussed in the next chapter.

6.7 Conclusion

In this chapter, building on the analysis conducted in the previous chapter, I have described the implications for an automobile company of not achieving the structure and size put forward by the automobile industry paradigm. I have examined various ways in which a company, uncompetitive within the automobile industry paradigm, can approximate to the paradigm. Internal approximations include extending production over time or intensifying production through flexible production techniques. However, neither strategy deals directly with the cost disadvantage of suboptimal output, although they may ameliorate the effects.

External approximations to the paradigm involve making alliances with other firms. Although intermediate forms of governance have been suggested as sufficient compromises in the make-or-buy dichotomy they do not represent novel forms of governance. This is because they are not in accordance with the possibility of divisibility in the full-function model and thus are simply adjuncts to it.

This chapter has sought to demonstrate that integrated companies are divisible in two dimensions. The first is the distinction between human and physical assets, as illustrated by the link and scale alliance structures. Scale alliances tend to be secured by physical asset specificity which has long-term applicability, whereas link alliances are generally seen to be subject to *ex post* opportunism and so only last for relatively short-term periods. If the link alliance is made between complete, discrete functions, however, then the advantages of new developments accrue to the partner responsible for the function and so opportunism does not arise. This indicates that VJVs involving complete and complementary functions may be the basis of long-term sustainable partnerships. The second dimension is the distinction created by factor endowment differentiation between different international regions. This indicates that factor endowment differentiation may be enduring and provide an international division to an alliance. In combination, the IVJV has the potential to replicate competitively the structure and size prescribed by the automobile industry paradigm.

Against the background of the automobile industry paradigm proposed in this book, in the next chapter I will examine the activities and subsequent collapse of the case study company, MG Rover. In particular, from the perspective of approximating to the automobile industry paradigm, this will focus on the company's attempts to secure its future, ultimately unsuccessfully, by way of international joint ventures, and how this opened the door to the Chinese automobile industry.

CHAPTER 7

Opening the door to China

I have set out the conditions for an automobile industry paradigm. This defines the theoretical structure for a prototypical automobile manufacturer in terms of scale and vertical integration. These two structuring forces have been assessed in terms of economies of scale and transaction costs, respectively. The resulting model has extended the Budd Paradigm to show how an optimal firm would integrate the four main processes (R&D, BIW, powertrain, and final assembly) into a full-function structure, each function exploiting the available economies of scale.

Survivor analysis has found that although the industry-leading automobile manufacturers benefit from strategic advantages, the MES for the industry is likely to be somewhat lower. A company that is close to the MES for the industry represents the prototypical form of company, and its output is signified by an MES_P at around 600,000 units a year. The onus is then on those companies that are suboptimal to the MES_P to formulate a strategic response that enables them to approximate to the advantages of the paradigm. As was examined in Chapter 6, this can be done internally, using the company's own resources, or externally, by accessing those of another company. Internal approximations tend to focus on the utilization of the physical assets by either stretching the production life of the product or exploiting economies of scope and using flexible manufacturing to intensify the production system. However, neither approach excuses the company from the cost implications related to MEPS or overall minimum efficient scale (MES_P).

External approximations to the paradigm do not include transacting with companies in the market, since this is diametrically opposed to the full-function feature of the automobile paradigm, not an approximation of it. In order to replicate the full-function structure by using an external mechanism the company must partner itself with another company in a manner that offers an alternative to the make-or-buy dichotomy. Alliances have been put forward as a quasi-internalized organizational structure and this can involve sharing of facilities in a horizontal joint venture. However, it was suggested in the previous chapter that joint ventures are most stable when they coincide with the divisible company activities. Since companies are most readily divisible along functionary demarcations I have put forward the concept of the VJV. This replicates the full-function model so that partner companies take responsibility for one or more of the complementary functions.

International factor differentiation also encourages divisibility by offering cost advantages when structuring around human and physical assets. Allied to the VJV structure, the resulting IVJV allows companies to continue to specialize in the human and physical factor endowments of their existing functions to their mutual benefit. If the functions are also competitive in terms of scale then there will be sufficient approximation to the automobile industry paradigm for the combined companies to be competitive.

In this chapter I will apply these considerations to a reinterpretation of the vicissitudes suffered by the British automotive company, MG Rover, during its five years of operation until its destruction brought to an end, for the time being, the era of British-owned mass-market car manufacturers. I will suggest a structure that might have prevented the company's ultimate demise and compare this structure to the final, posthumous structure of the company's assets. While this hints at a long-term sustainable solution to surviving the overwhelming tide of globalization, it also suggests an entry point for China into the global automobile industry.

7.1 The rise and fall of British Leyland/Rover Group

This section will present the historical antecedents to the independence of MG Rover in 2000. This commences with a brief history of the two brands, MG and Rover, describing how they came together. I will then establish the roots of the problems facing British Leyland (BL), and then Rover, in particular their failure to restructure, which led to a vicious cycle of lack of funds preventing new product development and thence further declining revenues. BL first attempted to counter this by improving its internal capabilities with injections of government funding. When this proved insufficient, it then turned to external assistance in the form of a joint venture with Honda, followed by ownership and funding by BMW.

The consolidation of MG and Rover

The MG Rover brand was comprised of two marques, MG and Rover, plus a variety of moribund brands. MG was founded sometime in the early

1920s under the guidance of Cecil Kimber, the exact date being unknown. It was owned by William Morris as part of his Morris Garages business until it was merged into the Morris Motors car manufacturing business in 1935. Morris Motors, as part of the Nuffield Organization (including Wolseley and Riley), then went through a series of mergers itself:

- 1952: merger with Austin to form the British Motor Corporation (BMC).
- 1966: BMC merged with Pressed Steel and Jaguar to form British Motor Holdings (BMH).
- 1968: BMH merged with Leyland to form British Leyland Motor Corporation (BLMC).
- 1975: the British government took a majority equity holding in BLMC and effectively nationalized the company.

Throughout this period the bloodline of the MG marque became increasingly diluted. The marque has always been a "badge engineering" marketing tool of its parent and has never had its own dedicated R&D or powertrain operations. It did have its own factory though, at Abingdon, from 1929 to 1980. After the factory closed the nadir was purportedly reached when the MG brand was attached to the poorly received Metro, Maestro, and Montego variants (Williams *et al.* 1987; Sharratt 2000). MG regained its credibility in 1998 with the release of the MG-F sports car, and still more from 2000 under MG Rover when attachment of the badge to sportier versions of Rover models was greeted more positively than previous badge engineering strategies.

Rover has an altogether weightier presence within the history of the British automotive industry, as documented by Graham Robson (1988). James Starley and Josiah Turner began The Coventry Sewing Machine Company in 1861, but the latest innovation in personal transport, the bicycle, prompted a change in direction in 1869. Starley then left to start his own company with cyclist William Sutton, using the name Rover, and producing the first safety bicycle in 1885. The company moved progressively into car manufacturing during the early years of the twentieth century and concentrated solely on cars from 1924. With the arrival of the Wilkes brothers in 1929 the products gained a reputation for solid reliability. However, the company also showed considerable innovation, such as the Land Rover off-road vehicle in 1947, the initial designs of which were based unofficially on the Willys Jeep, and the first gas turbine road car, JET 1, in 1950.

Like MG, the Rover marque was subsumed into successive mergers:

- 1967: merged with Leyland Motor Corporation.
- 1968: Leyland merged with BMH to form BLMC, renamed BL from 1977.
- 1976: Rover SD1 established an internal merger between Rover and Triumph.

After the launch of the SD1, the Rover brand became part of BL's marketing strategy, being applied to a variety of models, some of which had not started out as Rovers. By 1986 the Rover brand was chosen for the car-making group, changing from BL to Rover Group at the same time as Land Rover was spun off as an autonomous business unit.

Internal rescue strategies: 1975 to 1978

In 1975, BLMC had come to dominate the domestic automotive industry in terms of output, but it lacked internal resources to fund the replacement of aging and unprofitable product lines while still consolidating its operations after decades of mergers (Pilkington 1999). The combined operation comprised almost every aspect of the automotive industry in the UK, from trucks and buses to cars, but without restructuring it was simply the same set of companies under a single corporate banner. In the words of Michael Edwardes, chairman of BL from 1977 to 1982: "BL represents a microcosm of the issues affecting the British industry as a whole" (Edwardes 1983, 9).

It was not a single full-function firm exploiting available economies of scale, as prescribed by the automobile industry paradigm, but an affiliation of small full-function firms, all of them suffering the same scale disadvantages they had before the mergers. In the mid-1970s, BLMC was comprised of eight companies and 50 marques. Manufacturing capabilities included the Longbridge (Austin) and Cowley (Morris) plant complexes, plus another seven final assembly plants handling production for Jaguar, Rover, and Triumph. Each of these plants had remained attached to its historical brands, creating inflexibility in production (Bhaskar 1979). When flexible production was introduced in the 1980s, with more automated manufacturing systems, the low level of sales simply exacerbated the problem of low capacity utilization with the burden of the higher fixed costs of the new capital investment (Williams *et al.* 1994).

The merger that created BLMC had not come about through commercial necessity but political motivation. Using the mechanism of the Industrial

Reorganization Corporation, a government body was created to reinvigorate British industry, the elements of BLMC were forged together by Anthony Wedgewood Benn in 1968 (Church 1994). The government had been concerned about the inroads being made by US manufacturers, such as Chrysler's purchase of Rootes, and the national balance of payments. As the elements of the company were brought together it became apparent that there were not the managerial resources to cope with the new structure (Bhaskar 1979). The inclusion of the Leyland truck company was intended by the government to improve financial controls in the car manufacturing division. Pricing of cars at BMH, for example, was made without reference to the innovations on which the competitive advantage was founded: Ford calculated that the technically advanced Mini, then priced at £496 (US$1,190), lost £30 (US$72) on each one sold (Church 1994). Investment funds for the future were further dissipated by the company policy of continuing to pay dividends, even until 1975.

Some fault was found with the technicalities of production, Lewchuck (1987) criticizing the company for failing to adopt the Ford style of mass-production while retaining old systems of payment and control. This view did not find favor with Williams *et al.* (1994, 137), who pointed out that production engineer Frank Woollard had a good understanding of the need for ATMs and used precellular flow in a pull-through production style reminiscent of Toyota's TPS production system. This resulted in car assembly at Longbridge CAB1 (car assembly building 1) of 12.5 finished cars per worker. However, while Woollard understood the global innovations in production, output capacity did not find support in sales as it had at Toyota. Foreign opportunities were reduced by the effects of national protectionism, the company having little presence within the EEC while the home market was maturing. In 1975 imports took 33 percent of the UK market, overtaking BL on 31 percent for the first time, and the situation was exacerbated by the growing influence of North Sea oil revenues on the strength of the British currency. A crucial factor at this time would have been effective cost recovery to fund future investment, but instead £103 (US$229) was being lost per vehicle in 1975, deepening to £662 (US$1,540) by 1980 (Williams *et al.* 1994). The company produced 1.26 million vehicles in 1975, down from 1.7 million in 1973, and lost £76 million (US$169 million) (Church 1994).

Government intervention continued throughout the 1970s. Three reports were published during 1974–75: the Ryder Committee, the Trade and Industry Sub-Committee, and the Central Policy Review Staff. Most prominent was the 1975 Ryder Report, stating that the company suffered from outdated facilities, a weak organizational structure, and poor planning in consolidating its disparate operations (Álvarez Gil and González de la Fé 1999). The report was devised under the chairmanship of Don Ryder of the publishers Reed International (Williams *et al.* 1994, 151). Emphasis was placed on economies of scale, with the stated intention being to expand output to fill available capacity, rather than rationalizing the internal capabilities of the company. Bhaskar (1979) noted that BL was in short supply of engineers of all types, quantifying this deficit in thousands rather than hundreds. The Ryder Report suggested consolidating group R&D at one location, but this was resisted by Jaguar which attempted to protect its independent spirit.

The production funding entailed an investment of £2 billion (US$4.4 billion) in new models and capital equipment with a planned rise in production from 605,000 in 1975 to 961,000 by 1985, closer to the theoretical production capacity limit of 1.2 million (Williams *et al.* 1987). The Ryder Plan was criticized for being expansionist without addressing internal problems of a complex corporate structure, inadequate technical resources, and poor industrial relations (IR) of the company, which lost it 252,000 units in 1977 alone (MIRU 1988). The rise in production capacity might have been appropriate if the mass market had not shifted in favor of the business user, company cars having become established as an employment incentive during a period of pay restraint. Ford had come to dominate the business market since the 1960s, taking 40 percent of the segment by 1975 (Church 1994). BL traded on its product innovation which appealed to the private purchaser, but just as this segment was shrinking the sales were squeezed further by the rise in popularity of imports (Williams *et al.* 1987).

Bhaskar (1979) concluded that the Ryder plan had failed partly because it was ambitious beyond the competence of those enacting it, but the most important failing was in linking funding to improvements in IR and output. When stoppages brought down production, the funding that was available was spent on operational needs, instead of investment. From 1975 to 1977, BL continued to suffer falling production and market share, the car division alone dropping its market share from 32 percent in 1975 to 24 percent in 1977 (MIRU 1988). In 1977, Michael Edwardes was brought in from the National Enterprise Board (NEB), having turned round the battery maker, Chloride. Internal restructuring in 1978 meant that BL was reorganized with the car division being divided into Austin-Morris and Jaguar-Rover-Triumph and each accorded decentralized decision making. The Corporate Recovery Programme was presented to the government in July 1979 with support from the workers. It was based on a reduction in production capacity, from the 1.2 million theoretical limit to a planned 750,000 units a year, and a reassertion of the management's right to manage the firm. Michael Edwardes publicly stated that this strategy was "shrinking the business to a level beyond which we believed it was not safe to go" (Williams *et al.* 1987, 95).

Plants, such as Speke, were closed, and car production was concentrated on plants at Cowley and Longbridge. Government funding was not contingent on results; rather it was made available for investment in future programs. However, the company found that redundancies still lagged behind falling sales. Sales had halved to 315,000 between 1972 and 1980, but the workforce only halved to 55,000 in the period 1979 to 1983 (Williams *et al.* 1994). Moreover, the company continued to be dogged by poor economic conditions compounded by a lack of new models. With the company's internal capabilities still uncompetitive, Michael Edwardes decided that the salvation of BL cars lay externally, and so the search began for a JV.

External rescue strategy: link and scale alliances

Michael Edwardes stated that the partner had to be in a position of commercial coincidence: similar but complementary needs and aspirations within the same segments as BL (Edwardes 1983). The purpose of the JV was to improve BL's internal difficulties with new product development, the intention being to return the company to health as an independent and competitive automobile manufacturer. In terms of my research the need was for a JV which would return BL to a competitive status within the automobile industry paradigm.

The fear was that the company would lose its full-function status and become an assembler of vehicles, thus losing the political support that sustained it. Potential problems were identified as cultural prejudice against foreigners, language barriers, geographical distance, and an insecure future for the company if the partner firm dominated. Renault was looked at in 1978 and then rejected on fears that being larger and state-owned it had the resources ultimately to overwhelm BL. In the same year, Bob Price, managing director of Vauxhall, proposed that he could solve BL's problems by his company providing the mid-range replacement for the Marina, Maxi, Princess, and Allegro and receiving in return the small car design that Vauxhall lacked. However, the Gemini project being offered by Vauxhall was then canceled by its owners, GM (Edwardes 1983).

A proposal by Edwardes, codenamed "Dovetail," to solve the problem of poor R&D engineering resources at BL was to merge with Chrysler UK, close two of the Chrysler plants, and absorb its R&D facilities. It should be noted that it was the inability of the government to let Chrysler UK fail that had undermined attempts by BL to achieve market growth (Church 1994), the British government having a policy of supporting the subsidiaries of US firms as if they were indigenous even though they operated without reference to the United Kingdom: "their corporate strategies are international, their links are within the group rather than within the country, the major decision loci are abroad, they can survive without Britain" (Wilks 1990, 175).

Edwardes seemed to recognize implicitly that developing a link alliance in R&D would not have been stable and so a complete merger with Chrysler's R&D was the preferred option. The deal would have secured access to new engineering capability and one dedicated to new BL models, rather than one distracted by its own model range – this had been the very problem that had dogged the mergers that comprised BL and its antecedents. However, the merger was overtaken by the purchase of Chrysler UK by Peugeot in 1978. Other plans were to offload Leyland trucks onto Fiat/Iveco, yet another to have John Delorean take control of Range Rover sales in the US; but none of these came to fruition (Edwardes 1983).

In 1978, ex-ambassador to Japan, Sir Fred Warner, met Honda President Kiyoshi Kawashima. At that time, although the two companies were superficially competitive within the automobile industry paradigm, the two had distinct differences. BL was still handicapped by its multicompany heritage of 16 model families, 30 factories, and eight final assembly plants, and it had come under state control. Honda was a tightly structured private company with two model families and two assembly plants. There was little commonality in engine technology, Honda having promoted the CVCC (compound vortex controlled combustion) technology while BL had inherited whole families of power units. However, there were also the commercial coincidences that Edwardes was seeking. Production levels were broadly similar, British Leyland producing 800,000 cars and trucks to Honda's 740,000 (Mair 1994), figures that include SUVs, and they both specialized in front-wheel drive technology on cars in comparable market segments. This suggested that the JV would be complementary in engine technology, where BL could gain from Honda, but with commercial coincidence in the basic vehicle configuration, which would permit the companies to share in vehicle development. With equivalent levels of output, and Honda having no government funding to draw on, there would be little danger of the Japanese company absorbing BL.

The two companies also complemented each other in the US market, Honda being present with a small car and BL having the large Jaguar sedan. Initially, though, the first priority of the British company was to find a stopgap product to keep the production lines running between the launch of the Metro supermini in 1981 and the medium-size Maestro in

1983. In 1979, the "Co-operation for Business" deal was signed at the same time as the Corporate Recovery Programme reduced British Leyland to 13 factories of all types and shed 25,000 jobs. The Metro began the "product led recovery" in 1981 and in 1982 the Honda Ballade emerged from Longbridge as the Triumph Acclaim, built up from CKD kits. While it was not the first relationship between Japanese and British manufacturers (Nissan had built Austins under license in the 1930s and 1950s), it was, perhaps, the moment the Japanese industry came of age. It signaled the swansong of Triumph and the arrival of the Japanese as potential saviors of the British car industry.

Honda was still a relatively new car manufacturer in 1978. It had begun manufacturing cars in 1962 with the S360 sports car, but did not break into the mass market until 1969 when the Civic was designed for the US market. This was crucial to the future growth of the company: "we all believed that if the project failed Honda would have to give up its plan of becoming a full-fledged carmaker" (Hiroshi Kazawa, engineer, Honda 1999, 109).

The larger Accord built on the early sales penetration of the US market by the Civic and eventually went on to become one of the biggest selling passenger cars in the market. The company was also experienced in foreign production, particularly in motorcycles, where it had been manufacturing in Belgium since 1963 (Honda 2006). However, in Europe the company had no automobile production and little sales exposure, though the products were conceptually similar to European models (Mair 1994). Honda was therefore looking for two benefits from a relationship with BL: (i) market penetration; and (ii) production facility.

Categorizing the relationship according to the typology of Dussauge *et al.* (2004), from Honda's point of view it was predominantly a scale alliance, offering the company access to BL's final assembly facility. In the process, it could gain market knowledge and build contacts with European suppliers, though this could be achieved independently of BL. Figure 7.1 shows how the JV for the Triumph Acclaim brought together the assembly functions of the two companies' structures. Since the relationship involved assembly from Honda-supplied CKD kits, Figure 7.1 shows that BL's capabilities in BIW, R&D, and powertrain were largely irrelevant, although they were being employed elsewhere within BL. Furthermore, the market intelligence from the UK operation related to this model accrued mostly to the benefit of Honda in Japan.

For BL, Figure 7.1 shows that the relationship was a scale alliance in final assembly, allowing the company to continue operating an existing assembly facility until the new models came on stream from 1983, with

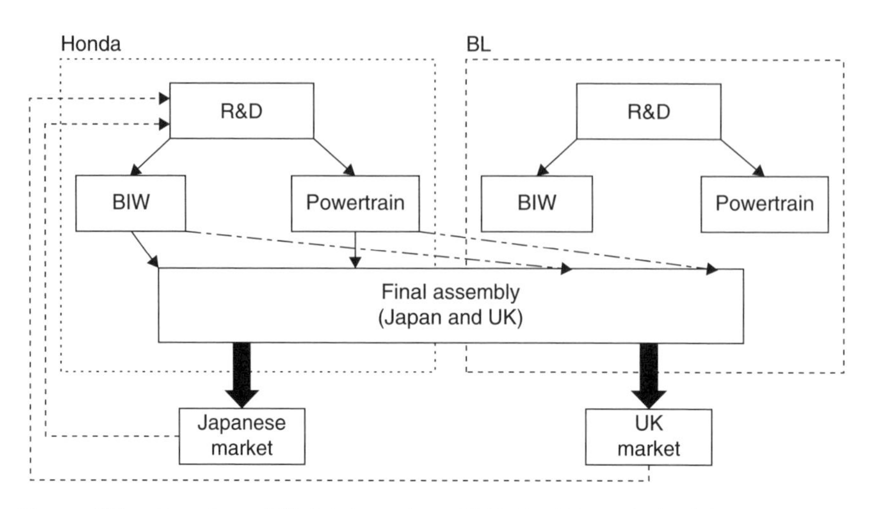

Figure 7.1 Honda and BL joint venture 1981: the Triumph Acclaim

the Maestro and then the Montego. Pilkington (1999) found little of financial or technical benefit to BL in the first stage of the relationship, its other functions being effectively isolated, but at least it improved the company's reputation in the market. Pilkington also states that it was from this point that BL recognized the potential for a link alliance, learning from Honda in order to improve BL's own R&D capability. However, Pilkington did not find any evidence that sufficient knowledge was transferred during the period of the Triumph Acclaim project.

The link alliance aspect of the relationship seems to have developed in the volatile manner that Dussauge *et al.* describe. Pilkington (1999) describes five major programs and the degree of input by BL. The first, the Triumph Acclaim, was a lightly reworked Honda Ballade, but it created the cooperative environment that led into the Rover 200 of 1984, which offered a single BL powertrain system. That aside, the Rover 200 was still heavily based on an existing Honda, the new Ballade, but its release in 1984 was actually preceded by the start of a deeper link alliance between the two companies.

In 1981, Project XX was announced, a new departure for both companies since it was the first large car for Honda and the first front-wheel drive large car for BL. By agreeing on common design references (or "hard points") the two firms could offer their own styling and powertrain variants. The final Honda product, the Legend, hid any British influence very comprehensively and no British engines were offered. This revealed Honda to be the dominant partner, as noted by Pilkington (1999). Indeed,

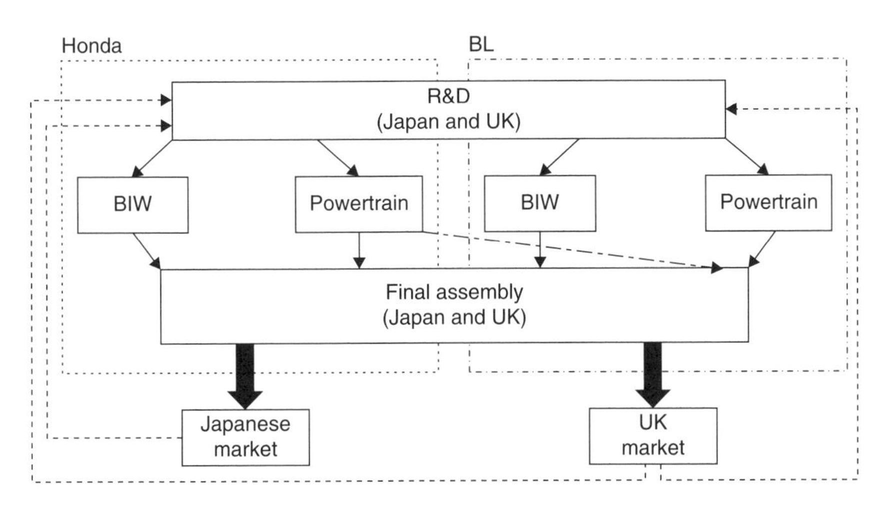

Figure 7.2 Honda and BL link and scale alliance 1987: Project XX

the Japanese seemed to specify the design to the degree that the only large engine that would fit the range was the Honda V6. However, the Rover 800 range was considerably more expansive, having four-cylinder British engines and a hatchback version in addition to the luxury saloon with the Honda engine.

Figure 7.2 shows the state of the alliance between BL and Honda during Project XX, with a link alliance in R&D and scale alliances in final assembly (both brands being assembled in both countries), but with only Honda providing engines to both versions. It also shows that Honda continued to benefit in the form of market intelligence from production exposure to the UK market.

The independence of the two companies on final execution of the vehicles was underlined when Honda was able to unveil their version of Project XX in October, 1985, a full nine months before BL showed theirs (Robson 1988). From the disjointed chronology it seems clear that Honda gained early from the link alliance and was then able to pursue its own model agenda. In the sense of being able to get to the market sooner the advantages accrued asymmetrically to Honda. However, despite the cost savings in sharing the design stage, it was still a prolonged gestation period, the later commencing Accord, a wholly Honda design, reaching the market before the alliance generated Legend (MIRU 1988). There were also plans for joint manufacture but Honda was reluctant to subject itself to BL's quality control, a policy vindicated in the statement by Williams *et al.* (1994) that even in 1991 the Rover 800 was averaging £325 (US$574) of warranty work per vehicle, four times that of the Legend.

In addition to forming a link alliance in R&D, BL needed to reform its structure and its IR policies (Pilkington 1999). With Honda's skills in manufacturing it might be assumed that this would have been achieved through use of the Japanese production technology, but in fact only 10–15 percent of BL's subsequent improvement in productivity was due to technological innovation (Scarbrough and Terry 1996). Mair (1994) even asserted that Nissan was more influential because it set the industry standard, though BL would later gain more by easy access to Honda's experience in starting the Ohio plant, rather than by Honda directly instructing BL in the UK. At the time of the commencement of the alliance the reputation of Japanese manufacturing was low, the Japanese advantages believed to be derived from the factor differentiation of low-cost labor, exploitation of suppliers, and copying of Western technology. Initially there was organizational resistance within BL to learning from Honda, and British visitors to Japan were only learning by observation, the language barrier making the culture impenetrable to them. The Honda experience in Ohio should have been an invaluable opportunity for BL, but it was not until 1991 that the British started their first formal studies of the Honda culture. As described by a BL manufacturing manager: "we had our share of problems with Honda to start with. Then, after a while, it all sank in. Why all the aggravation? They're right, we're wrong. All of a sudden the concept changes, and things fall into place" (quoted in Mair 1994, 273).

BL did not adopt Honda's method of production in batches, which was a US market-oriented build-to-stock system requiring extended planning (Pilkington 1999). Rather, BL had its own minimum inventory control/strategy (MIC/MIS) system using a version of JIT but with large central warehouses and a type of *kanban* inventory flow control system which was, however, not as flexible (MIRU 1988; Pilkington 1999). Furthermore, the production system was purchased from Honda and thus did nothing to improve BL's production engineering capability. Although the joint venture was intended to be a link alliance, transferring knowledge to BL by accepting Honda's control over product and production engineering, BL's own capabilities were being emaciated: "it is all too clear that this attraction [to accept Honda systems] diverted the firm away from implementing a learning-based strategy" (Pilkington 1999, 467).

BL was in a position, though, to use the period of stability provided by the alliance to conduct some reorganization. It was also fortunate that the fundamentally perilous state of the company had created an understanding by the unions of the need for reorganization. The company asserted the right to manage, which found expression in the direct communication with the labor force, often from Michael Edwardes himself. This marginalized the unions, but also built trust and cooperation with the workforce, based on the particular circumstances of the company at that time. It was not, therefore, something that could simply have been imported from Honda. Indeed, as Mair (1994) stated, BL was learning from Nissan as much as from Honda. The reorganization of IR was started before the relationship with Honda was promulgated, commencing with the teamwork system for the all-British Metro, although Smith (1991) claims that this was a case of workers coping and looking out for each other. In 1987 a TQ (total quality) initiative was introduced and further refined for "Rover Tomorrow – The New Deal" in 1991. This was formulated by senior management, being only indirectly influenced by experience with Honda (Scarbrough and Terry 1996). This recognized the poor existing economies of scope in management: "the 'enemy' isn't the Japanese or the Germans; it is the limitations of our managerial 'culture'" (Pascale and Athos 1986, 201).

From 1982 BL looked to capitalize on the gains made from the link alliance with Honda by restructuring the company under Sir Austin Bide into the Cars Group and Land Rover–Leyland Group. Jaguar was privatized in 1984, and in 1986 an attempt was made to sell BL to Ford, but this was considered politically unacceptable in the wake of the fallout from the scandals involving Westland Helicopters. In the same year a record loss of £539.2 million (US$791.5 million) was announced and BL was renamed Rover Group when Graham Day was installed as the new chairman. In 1987 the group was further broken up with the privatization of Unipart, Leyland, and Freight Rover. Recognizing that the remaining car company, Rover Group, was still not ready for independence, Day called for recovery in three phases, involving restructuring, then restoration of Rover Group to commercial viability and finally privatization (MIRU 1988). However, the four alternative recovery strategies that were under consideration seemed to recognize implicitly that the company could not make a case for independence as a mass-market manufacturer. The four options were:

1. Selling Rover to another manufacturer.
2. Selling the group to Honda.
3. Moving to low volume specialist production with premium pricing.
4. Selling the group to a non-automotive manufacturer.

Since the first three options implied that Rover Group needed a partner within the industry, the fourth option appeared to have little logic. Although this was indeed the option that came to pass, it was only effected with the cooperation of Honda.

BAe purchases Rover Group

British Aerospace (BAe) first showed an interest in purchasing Land Rover in November, 1987, to provide an add-on to sales of military aircraft. Instead, in 1988, Rover Group in its entirety was sold to BAe for £150 million (US$267 million) with £520 million (US$926 million) in government aid and tax allowances (MIRU 1988; Church 1994). The automobile alliance was secured by equity cross-holdings, Honda taking a 20 percent share in Rover Group, while Rover Group took a 20 percent share in Honda UK Manufacturing (HUM) in 1989 (Mair 1994). This indicated that despite the fluctuations in the degree of collaboration for successive projects, Honda was prepared to take a long-term view of the JV.

This politically acceptable solution provided long-term stability for Rover with minimal disruption to suppliers and Honda. It was mooted that the two British companies would be able to use a common computer aided design (CAD) system and merge administrative operations in order to benefit from synergy. As it was, the relationship with Honda had been continuing with the release of the Rover 200/Honda Ballade in 1984, providing valuable assembly work for Rover. The Ballades made by Rover numbered only 30–50,000 per year, but this was a significant utilization of production capacity and with 50 percent of the value-added being British it made a further contribution to the domestic industry as a whole (Williams *et al.* 1987). This scale aspect of the alliance lasted for the duration of the model (Pilkington 1999). However, Rover was still unable to extend the link aspect of the relationship into making an entrance into the US market under Honda guidance, a move that was further undermined by Day's insistence on withdrawing from markets that did not provide acceptable financial returns. The company was caught in a vicious cycle of not investing in markets where there were poor sales, with the lack of investment then leading to a further deterioration in market share. Despite the scale alliance aspect of the relationship with Honda, the continuing lack of export sales meant that the three assembly plants in the UK were not being fully utilized. According to Williams *et al.* (1987) the plants at Longbridge, Cowley South, and Cowley North had a combined production capacity of 780,000 units in 1987, but only 469,000 units actually left the factory gates that year. This equated to a production utilization of around 60 percent overall.

Throughout the period under BAe ownership Rover Group benefited very little from its parent. Financial stability allowed the alliance with Honda to progress but then Rover Group had to forgo its involvement in successive projects when a lack of funds forced it to instead extend production of existing models, albeit with refreshed styling. Indeed, Rover Group was at the mercy of BAe's financial status, and the economic recession during the ownership period (1988–94) reduced the funding available to the car operation (Pilkington 1999). By this stage in the link alliance Honda had learned as much as it required from Rover Group about UK production and had expanded its assembly plant in Swindon. Rover Group, however, had been frustrated in its requirement for knowledge transfer from Honda. Each of the joint five projects had been to Honda's specification, leaving Rover Group to make adaptations. The pinnacle of the relationship was Project XX, but Rover Group lacked the funds to join Honda in its replacement in 1991 and so was forced to extend production of the Rover 800 version, an unsatisfactory strategy in a globalized industry (see Chapter 6). The final joint project, resulting in the Rover 600, had very little Rover involvement and the company was even reduced to seeking assistance from external agencies in adapting its production facilities to the new model (ibid.).

However, although Rover Group was not gaining as much from direct involvement with Honda as critics suggested it should, it was an opportunity to concentrate what resources it had on its own projects. The company designed and installed its new range of petrol engines, the award winning K Series, comprising the full range of small in-line fours to larger V6 engines, available from 1988. The company had also retained its ability to conduct new model programs by itself, fundamentally re-engineering the Metro into the Rover 100 and delivering on the Land Rover Discovery and Freelander (ibid.). The Freelander was an entirely new venture for the company, combining the rugged capability of Land Rover with the Budd Paradigm BIW production techniques of the passenger car division. The vehicle also, therefore, proved that Rover was still capable of production engineering. As Álvarez Gil and González de la Fé (1999, 425) stated concerning Rover's R&D capability during the period: "it has maintained the core skills needed to develop new vehicles, like the Rover 100, but it does not have the financial resources to maintain vehicle development across the full range."

The situation in 1994 was that Rover Group had consolidated around three assembly plants, had its own powertrain and BIW facilities, as well as a capable R&D operation. It was therefore a full-function company handicapped by poor levels of output which reduced its ability to make funds available for investment. Having already exhausted its internally generated attempts to reorganize sufficiently during the Ryder plan and Edwardes eras, the company had a continuing need for another external relationship, particularly if this came in the form of funding.

External assistance for internal improvement: BMW

By 1994, from Honda's point of view, the link alliance was reaching maturity (Carr 1999). The company had its own manufacturing facility at Swindon, producing engines since 1989 and complete automobiles from 1992 (Mair 1994). The company also handled its own European sales distribution and marketing and had established its regional headquarters in Slough in 1989 (Honda 2006). Although it was not full-function in the UK, with all R&D taking place in Japan, it was able to operate quite independently of Rover. This may have contributed to a sense of detachment from Rover Group, and despite BAe suffering financial problems with its regional aircraft division the decision to sell Rover Group in 1993 appears to have taken Honda by surprise, despite it being a known fact that BAe's agreement with the UK government not to sell Rover Group before 1993 was coming to an end (Bardgett 2000).

Honda suggested raising its equity share to 47.5 percent if BAe retained an equal share and Rover Group employees were granted the remaining 5 percent. According to Brady and Lorenz (2001) this would have valued the company at £650 million (US$976 million), so would have required an investment from Honda of around £179 million (US$269 million) for the 27.5 percent to add to the 20 percent it already owned. However, demonstrating that BAe gained little of operational value from Rover Group, the aircraft manufacturer was interested only in a complete sale to raise all available funds. This left two more options: a management buyout (MBO) with Honda retaining the 20 percent share, or acquisition by another company. Brady and Lorenz discount the MBO option because Rover needed $1 billion (£666 million) in working capital, which financiers were not prepared to lend.

The third option might have created a three-way partnership, Honda and another company both supporting Rover Group through a 20:80 equity partnership. According to Brady and Lorenz (2001), Honda was looking to enhance its leadership in joint R&D projects by expanding its managerial influence over Rover Group and to protect the flow of royalties paid by Rover Group for Honda designs. This it could not have done if its co-owner was another car maker with operational plans for Rover Group. When BMW was revealed as the interested party it therefore brought Honda's plans for Rover Group to a halt. BAe sold its 80 percent share in Rover Group to the German executive car manufacturer for £800 million (US$1,200 million) (Bardgett, 2000) and a month later Honda swapped its 20 percent holding in Rover Group for Rover Group's 20 percent holding in HUM. All future collaboration between Rover and Honda was abruptly canceled, though this did not affect existing models which continued in production, two even outliving the period of BMW ownership. In retrospect, Honda seems to have been relieved at extracting itself from the alliance with Rover Group (*Economist* 2000).

BMW considered that Rover Group lacked sufficient volume and saw the purchase of it as a route into the mass market without risking the premium image of its own brand. The alternative was to create a new brand of its own, but Brady and Lorenz (2001) point out that such a development can be an expensive strategy, quoting a figure of $5 billion (£3.3 billion) for Toyota to establish the Lexus brand. For Rover Group, the strategy of protecting the BMW brand by keeping Rover at a distance had substantial benefits for its engineering infrastructure: "this rendered Rover substantial autonomy, but compromised integration and scale economies, and the strategy then depended heavily on some revived basic engineering capability" (Carr 1999, 417).

If "integration" is considered to be analogous to a link alliance, then without that close connection or the exploitation of economies of scale either, it is difficult to conceptualize BMW's involvement as little other than a source of investment funds. There was speculation that BMW's main interest had been to use the advantages of ownership to secure access to Land Rover's off-road technology. To a certain extent this seems to have been effective, providing the German company with an opportunity to compare technologies: later BMW off-roaders featured the same type of "hill descent" mechanism as Land Rover. However, it is not clear to what degree this was a Land Rover development, and it is even available on Jeep models that have no connection to BMW or Land Rover. Furthermore, the BMW X5 that emerged after the company divested itself of Rover owed more to BMW's own 5 Series platform and it used a development of the four-wheel drive system that had been in the company repertoire since before the purchase of Rover Group. Concerning the new Mini, the vehicle was proposed by BMW, not its Rover subsidiary, even though the style clearly paid homage to the British original.

Rover itself needed to design a new range of vehicles in place of those that presumably would have come out of the alliance with Honda, and this involved a complete restructuring of the Rover Group's R&D capability. This necessarily resulted in a substantial delay and the only new Rover model to emerge during BMW's tenure was the Rover 75, released in 2000, and having little in common with any BMW model. Some commentators believe that the front-wheel-drive 75 was based on a rear-wheel-drive BMW 5 Series platform, mainly due to the existence of a transmission tunnel down the center of the Rover's floorpan, though MG Rover later found when developing the rear-wheel-drive version of the 75 that the rear of the floorpan had to be adapted just to make room for a differential.

At least the electrical system for the 75 did have a BMW base, which then caused problems for MG Rover when attempting to have it interface with the V8 engine sourced from Ford. There were indications that BMW was planning to increasingly integrate Rover into the BMW infrastructure, the new Range Rover being well advanced by the time Land Rover was sold to Ford in 2000 for £2 billion (US$3 billion) (Brady and Lorenz 2001), and on release in 2003 it was installed with BMW engines. If BMW had further subsumed the mainstream models of Rover Group strategically in this way it could conceivably have led to further delays in model releases, perhaps for the BMW branded vehicles as well.

Although BMW provided the investment funds that Rover Group's low output denied it, this still did nothing for output during the interim. From 1994–99, Rover Group's share (including Land Rover) of the passenger car market in the UK fell as the products that had come out of the JV with Honda steadily aged in the market. According to a government report (Bardgett 2000), the 12.8 percent market share in 1994 fell to just 6.5 percent in 1999. Financial losses deepened in parallel with the loss in sales, from a loss of £158 million (US$249 million) in 1995 to £750 million (US$1,213 million) in 1999. Rover Group management disputed the scale of the losses and even claimed a notional profit (Brady and Lorenz 2001).

The strength of the British currency at the time suppressed Rover Group's exports, while the company was slow to capitalize on the relative reduction in cost of imported parts, gradually raising the proportion of imported parts in the Rover 75 from 15 percent to 25–30 percent (ibid.). In 1999, BMW sought government aid for its British investments and was promised £152 million (US$246 million) in public funds on condition that it invested £3.3 billion (US$5.3 billion) to secure future UK production. Suspecting that this contravened European regulations on state aid, the EU Commission announced an investigation. This lasted for several months into 2000 (Bardgett 2000). When, in mid-March, 2000, BMW finally decided to cut its losses and divest itself of Rover and Land Rover it immediately felt the benefit as its share price rose 30 percent on the news (*Business Week* 2000). After decades of mergers and consolidation, Rover once again found itself independent.

7.2 MG Rover: autonomy and sustainability

During the disposal process Rover became the subject of a short bidding war between Alchemy Partners, led by John Moulton, and Phoenix Venture Holdings (PVH), owned by John Towers, Peter Beale, John Edwards, and Nick Stephenson through their financial vehicle, Techtronic (2000).

Alchemy Partners considered that Rover was not capable of supporting a three-model range of saloon cars, particularly as two were already outdated, and intended to build on the MG sports car with the Longbridge workforce reduced by half to 4,500 (Holweg and Oliver 2005). This created some political resistance in the UK, while BMW was unsettled by the possibility that under the Insolvency Act (1986) a deal with Alchemy would leave it exposed to Rover redundancy payments should the company fold within two years (Brady and Lorenz 2001). The PVH bid then came to the fore, comprising:

- A purchase price of £10 (US$15) (never actually paid).
- £500 million (US$758 million) in loans and stock from BMW (including a £427 million (US$647 million) loan), repayable by 2049 or, if Rover was sold, £275 million (US$417 million) immediately with a further £225 million (US$341 million) within 3 years.
- Production of 200,000 units a year.
- Redundancies of 2,000 personnel (from 8,800) at Longbridge.

Under the private ownership of PVH, MG Rover consolidated all functions at Longbridge. Rover's R&D facilities were split with Land Rover, though it was claimed that around 600 engineers were retained by MG Rover (*Automotive News* 2001). The company inherited the new Rover 75 and the Honda-derived 25 and 45, while MG was launched as the sports variant of the existing saloon range in addition to the MG-F sports car. Apart from body panel stamping, which stayed with BMW and so was sourced from the company's facility at Swindon, MG Rover had the basis of a full-function structure all situated at Longbridge. However, scale was another matter, and in this respect the company was substantially uncompetitive within the definition of the prototypical automobile company which prescribed output of 600,000 units a year and an R&D strength of around 1,200 engineers (see Chapter 3). In comparison to the MES_P, therefore, MG Rover had R&D at half the necessary strength and production capacity at a third of that deemed necessary by the paradigm. The company therefore needed to establish a strategy for approximating to the paradigm, either internally or externally, in a manner sufficient to ensure its sustainability.

Internal approximations to the paradigm at MG Rover

Internal approximations to the automobile industry paradigm include two possible approaches. The first is to extend the model production runs and so achieve high levels of capital use over a longer period (see Chapter 6).

As has been shown, lack of funds had previously forced the company to extend production runs and so miss out on joining Honda in successive new model programs. On independence, MG Rover had no complete new models ready for release, except for the estate/stationwagon version of the Rover 75 that had been developed, and canceled, under BMW. Costs of complete new model development has been estimated to be up to £1 billion (US$1.8 billion) (*Observer* 2004), although I have argued that annual R&D costs of continuing new model programs can be as low as £250 million a year (US$460 million at 2004 exchange rates) (see p. 65). Even the lower figure was beyond what the company could budget for, so it was obliged to continue with the models it inherited until it could find a strategy to introduce models of its own. In any case, extending model production runs is not an appropriate strategy in a globalized industry where manufacturers are in unfettered competition with one another, except perhaps when an established platform can be used as the basis for a new low cost variant (e.g. the Renault Logan).

Intensification of the production capacity at Longbridge offered the second alternative for approximating to the automobile industry paradigm. This would have meant introducing model variants to serve market niches, raising total output towards the target of 200,000 a year. By 2001 the company had a full line-up of MG variants of the three Rover models, although the rugged looking "soft-roader" version of the Rover 25, Streetwise, did not appear until 2003. In that same year two new models were released, the MG SV and the CityRover; but since both of these were manufactured elsewhere (Italy and India, respectively) they did nothing to intensify the use of Longbridge production facilities. Furthermore, despite the introduction of the MG variants, the slide in production output first experienced under BMW ownership continued with PVH (2000–05). Once again in its history the company suffered the effects of a strong currency and lack of global purchasing power (*Financial Times* 2002b), offset slightly by home market sales buoyed by a strong domestic economy. Table 7.1 shows the slide in production for complete years under both ownership regimes.

Although the data shows that the Longbridge installation was capable of producing the targeted output of 200,000 units a year (*Automotive News* 2001), the fall in production was so relentless that it is difficult to see how the trend could have been reversed without entirely new models.

Options for collaboration

According to Bailey (2003), under the ownership of BMW the factories at Cowley (producing the Rover 75 and, from 2000, the Mini) and

Table 7.1 Longbridge production output, 1999–2004

Year	Production output
1997	343,157
1998	281,390
1999	172,099
2000	146,757
2001	163,144
2002	147,037
2003	133,557
2004	106,293

Source: IHS Global Insight.

Solihull (producing the Land Rover) received investment in preference to the Longbridge plant, which was also manufacturing the original Mini. Longbridge facilities improved slightly when the Rover 75 production system was moved to the Longbridge site and the old Mini was phased out in 2000. However, Bailey concludes that the MG Rover models were aging, requiring new model development, either independently by MG Rover or with a partner. The alternative was a reduction in output to small-scale, niche sports car production. "For this reason, many industry experts do not believe that MG Rover has a viable long-term future" (Bailey 2003, 73).

The most pressing need, therefore, was to invest in R&D for a new model range. The only funds available were the £427 million (US$647 million) interest free loan from BMW, repayable by 2049, along with the estimated £350 million (US$530 million) value of the plant and stock (DTI/Auto Industry 2007). My research suggests that this might have been enough to inaugurate an R&D program for a few years but not enough to replace the entire product range.

Upon independence in 2000, MG Rover had embarked on an immediate search for an international partner. John Towers, the new CEO, openly discussed Chrysler and Honda as possibilities (*Automotive News* 2000), though these suggestions never rose above mere speculation. As production and break-even targets were missed the company announced a collaboration with a division of China Brilliance in March, 2002, to develop a new small car. Originally billed as central to the future of the company, by the time the deal started to go awry, just months later, the criticality of the deal was being downplayed. Concurrent deals were being sought with Tata in India for a possible small car and the Daewoo production facility in Poland as a possible bridgehead for Rover's entry into Eastern Europe and the establishment of a production base for export (*Automotive Engineer*

2004a). *Professional Engineering* (2003, 27) reported optimistically on the release of the CityRover, based on Tata's Indica, and the Streetwise, based on the Rover 25: "so if MG Rover is serious about wanting to grow, it may eventually have to attract and cede its autonomy to a major player. For the moment, it's all to play for."

Automotive Engineer (2004b) identified a joint venture with Proton as a possible candidate; but, like the Tata relationship and the initial collaboration with Honda, this involved the supply of an existing Proton model (the Wira) and using it as the basis of a new Rover. An attractive characteristic of the proposed alliance for MG Rover was that the two companies operated in complementary markets. The collaboration would have replaced the failed proposal for a deal with China Brilliance, which ended in 2003 with the imprisonment of the Chinese company's sponsor of the deal; but at least it did raise some consultancy fees for MG Rover (DTI/Auto Industry 2007). It was also mooted that since Proton was the parent company of Lotus, the sports car manufacturer and engineering consultant, that the two British companies could work together in the kind of link alliance described by Dussauge *et al.* (2004).

Another proposition known at the time (*Automotive Engineer* 2004b) was for MG Rover to take over the ex-Daewoo factory in Poland and install it with obsolete Rover 45 production hardware for sale of the model in East Europe. This also fell through in 2003 when MG Rover's bid was rejected (*Professional Engineering* 2003), though, in any case, there was no reason to expect the Polish market to be receptive to an aging design. Although the proposal suggested that the firm understood it needed to increase rapidly output capacity, and another plant could have taken it close to the MES_P, a jump in output volume of this magnitude would have been very risky for a firm of this size. This is a problem inherent to small companies: they need to expand output to achieve the MES_P, but they have to do so in steps that are each below the MEPS. Other rumored joint ventures included the licensing of engine production in India, advising Sonalika on the design of an off-roader, and the possibility of licensed assembly of the Rover 75 in a remote corner of Iran. None of these were of any significance, and indeed the Iranian deal appeared to be nothing more than self-promotion by the politician who oversaw the region.

The final possibility for an alliance came with Shanghai Automotive Industry Corporation (SAIC) in 2004, along with Nanjing Automobile Corporation (NAC) in a minority role. SAIC was in production JVs with VW and GM in China but it had no vehicle designs of its own, although it had just acquired the Korean SUV manufacturer Ssangyong (DTI/Auto Industry 2007). The Chinese company paid MG Rover £67 million (US$123 million) for the property rights to the Rover 25 and 75, along with the K Series engines (Holweg and Oliver 2005), but was reportedly annoyed by speculation that it was about to invest £1 billion (US$1.8 billion) in MG Rover (DTI/Auto Industry 2007), a figure that concords with some estimated costs of new model development – and which I argue would have sustained R&D for four years. The proposed deal foundered in March, 2005, concurrent with the financial collapse of MG Rover. Holweg and Oliver (2005) claim that SAIC's intention had been to gain access to MG Rover's intellectual property rights (IPR) and production equipment for use in its Chinese plants, which it would be better able to do if it bought them outright. However, against global competition, SAIC would have had the same pressing need to replace the model range as did MG Rover before it.

Holweg and Oliver (2005) conclude that MG Rover started out in 2000 with an aging product line that was suffering declining sales in a shrinking market segment and so lacked the incoming funds to invest in new models. This made the need for a JV partner critical to its long-term survival, and due to the production overcapacity in the West only partners in rapidly growing developing markets, such as China and India, would be suitable. However, it is my view that Holweg and Oliver, like many others, take too narrow a view of MG Rover as a source for partners of automobile models. They demonstrate MG Rover's need for a partner to bring salvation but do this so comprehensively that they leave no reason why a partner should be interested. With MG Rover apparently inadequate in every major area, from the model line-up to R&D investment, there did not appear to be anything that MG Rover could offer an interested party. On this basis the company was only worth the value of the land it stood on, much of which had been sold a year before, and the value of its more general-purpose production equipment. Certainly, after a pessimistic assessment of MG Rover's product range, it is difficult to see how any prospect would be attracted to the company's IPR. It is my view that the company did indeed have value as a contributing partner in an automobile manufacturing alliance, but only if it drastically reduced the scope of its activities.

Lack of scale in all its functions meant that MG Rover had little to offer in terms of production capacity, and indeed in the one partnership that it did form, with Tata – production took place in India even while Longbridge was operating well below capacity. Furthermore, the available capacity at Longbridge was relatively low at a declared 200,000 units a year, competitively close to the MEPS for one plant but a third of the MES_P, so a scale alliance would have necessitated investment in new facilities and a tripling in market demand. However, R&D was only half of the MEPS for the process and this could have been more rapidly expanded than could production

Table 7.2 MG Rover R&D expenditure, 2003–5

Year	R&D expenditure (£m)	R&D expenditure (% of sales)	R&D expenditure/ employee (£)
2002/3	45.5 (US$68.3 million)	2.7	6,800 (US$10,200)
2003/4	90.5 (US$148.0 million)	5.2	13,700 (US$22,400)
2004/5	23.9 (US$43.8 million)	1.4	3,700 (US$6,800)

Sources: DTI (2003, 2004, 2005).

output, since the core team of engineers was already in place. A link alliance, therefore, would have shown more promise, but only related to the transfer of knowledge concerning new models, not the existing range. A link alliance in R&D for MG Rover would have served the purpose of intensifying its R&D activity and thus improving output in that function.

In the case of MG Rover it is not possible to make a direct comparison with the figures for R&D since it did not release any new core models, although it did release a number of interesting variants of the range it inherited from Rover Group when it was sold by BMW in 2000. Table 7.2 presents the R&D results for the company as surveyed by the DTI for the periods 2002/3, 2003/4, and 2004/5. The peak period, 2003/4, was boosted by additional funding due to the evolving alliance with SAIC of China, while 2004/5 was cut short by the financial collapse at MG Rover. Each of these years is therefore a special case in its own right; nevertheless the disparity between R&D expenditure at MG Rover and the rest of the industry is of a magnitude so great that comparisons can still be drawn.

If 2003/4 is taken as indicative of a period during which new model R&D was progressing at an acceptable pace then it is instructive to compare this with other car firms at the same time. Table 7.3 does this for the same companies that were shown in Table 3.5 (period 2005/6), with the addition of MG Rover. While it shows that MG Rover R&D as related to sales and employee numbers was within the industry standards it only shows the rate of investment; it is the measure of the fixed cost requirement for new model programs that is more interesting. As was previously shown in Chapter 3 there is a great deal of disparity in absolute levels of R&D spending. Ford and Toyota's R&D budget for that year was up to 40 times larger than MG Rover's. These are the industry leaders in terms of total output and have much wider product ranges to develop, as well as a certain amount of pure research outside of the automobile paradigm as it is defined here (e.g. fuel cell research is unrelated to the powertrain function in the paradigm). Subaru was identified in Chapter 3 as indicative of the MEPS for the R&D function and it can be seen that MG Rover

Table 7.3 Global R&D expenditure, 2003–4

Company	R&D expenditure (£m)	R&D expenditure (% of sales)	R&D expenditure/ employee (£)
Ford	4189.7 (US$6,850.2)	4.5	12,800 (US$20,930)
Toyota	3484.0 (US$5696.3)	4.3	13,200 (US$21,600)
Subaru	313.3 (US$512.2)	4.4	11,400 (US$18,600)
MG Rover	90.5 (US$148.0)	5.2	13,700 (US$6,800)

Source: DTI (2004).

Table 7.4 MG Rover R&D productivity

Company	R&D staff	Output (millions)	Ratio staff:output
MG Rover (target output)	500	0.20	1:400
MG Rover (2004 actual output)	500	0.11	1:216
MG Rover–SAIC (combined target output)	700	1.00	1:1,429

Sources: *Financial Times* (2005), *Automotive News* (2006b).

was investing about a third of Subaru's financial commitment for the same year. Although this was well below the minimum required, around £250 million (US$460 million) according to this research, in absolute terms this amounted to a requirement for around another £160 million (US$293 million); compared to other firms in the industry, this was not a large sum.

Although by industry norms MG Rover's R&D staffing level compared poorly with output (see Table 7.4), the company enjoyed heritage benefits from the investment made by BMW and had also shown some ability to explore new projects. The proposed JV with SAIC would have resulted in an effective increase in output by gaining access to the Chinese plant and market (DTI/Auto Industry 2007), supported by an increase in R&D engineering staff. The staff numbers would have then been sufficient for the MEPS in R&D, while production output of 1 million units a year would have exceeded that which I have argued would be necessary for the MES_P.

These speculative figures taken from media text sources in Table 7.4 for MG Rover indicate the kind of productivity that was available in R&D. In such a scenario, MG Rover would use its product development function to support the brand and would look for a partner that was devoid of a product brand but could offer substantial production capabilities. This would have resulted in a link style alliance, but one where any *ex post* advantages in R&D would only accrue directly to MG Rover if it had primary responsibility in that function. This would only be possible if the partner had little or no R&D capability of its own but was competitive within the industry in terms of production scale, a scenario that describes a number of emerging

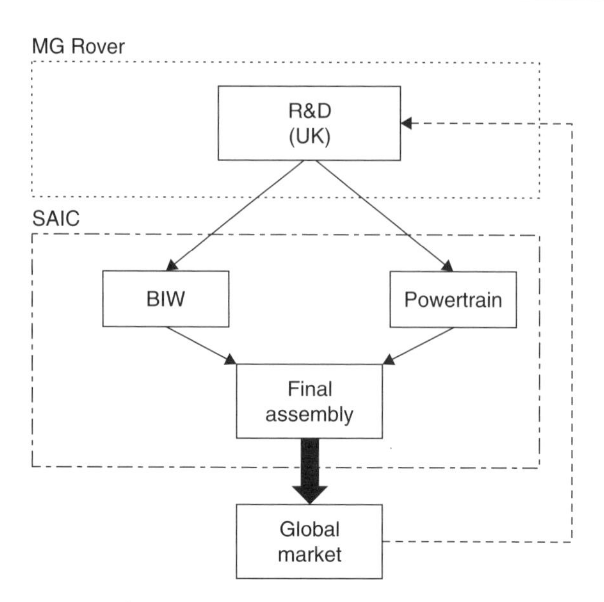

Figure 7.3 Proposed IVJV between MG Rover and SAIC

economy manufacturers. However, it also implies that significant production at Longbridge would have had to cease in favor of production facilities operated by the partner. With advantages in the functions accruing to the partner responsible, long-term stability could have been achieved. Furthermore, each partner would continue to exploit the international factor differentiation advantages of its home region. Figure 7.3 depicts how the resultant IVJV might have looked with SAIC.

The suggested closure of Longbridge as a production site recognizes that an organizational JV structure of this kind is not sustainable between companies that operate existing full-function structures. This precludes producers such as Proton or Tata that had attractive production capacity but also had product strategies of their own (*Automotive Engineer* 2004b). It is in accordance with this view that the purported alliances with these companies failed to develop.

Although the IVJV seems to have a solid theoretical basis, a significant problem concerns a suitable governance structure. I have found that a development of the U-Form of governance structure has tended to prevail over the M-Form, particularly amongst Japanese firms, but such a centralized form of management control would not be applicable between independent parties. The example of GM and A. O. Smith (see Chapter 6) suggested that the M-Form might be more appropriate since it describes the separation of strategy from operations. However, the IVJV is missing two crucial features: the transfer of objective data for strategic analysis and the existence of a designated headquarters. Neither of these factors can be taken for granted between independent parties without being explicitly implemented.

7.3 MG Rover assets: reconstructing the international vertical joint venture

In April, 2005, PricewaterhouseCoopers acted as official receivers to dispose of the assets of MG Rover. By July that year the process was largely complete, with the main physical assets and the MG brand being sold to NAC, previously the minor party in the proposed JV with SAIC. The lack of production location advantages in Longbridge were confirmed when NAC proceeded to extract the production equipment and move it to a new plant in China, restarting production in March, 2007. Although NAC had the physical means of production it did not own the product designs, which had been previously secured by SAIC during negotiations for its aborted IJV with MG Rover. NAC therefore had to reverse engineer the products in order to derive the technical designs. NAC found ownership advantages in the MG brand while SAIC attempted similarly to obtain the Rover brand name but without success, as Ford took up an option to purchase the brand from BMW in September, 2006.

The division of MG Rover between NAC and SAIC took in some of the elements of the OLI paradigm. The total investment by NAC and SAIC in acquiring MG Rover's assets in the UK amounted to £110 million (US$200 million) (DTI/Auto Industry 2007), so the companies clearly saw value in MG Rover. As owner of the physical assets, NAC seemed to perceive few locational advantages to the UK except for maintaining brand equity, which meant retaining a UK connection in the form of limited sports car production in order to provide some authenticity to the British brand image of MG (*Times* 2007). At a press conference in 2006, NAC made the MG brand ownership advantages very clear, but the locational advantages of the UK were more equivocal. Like most companies, NAC gave first priority to achieving production scale by bringing its primary plant in China to full capacity production. Stadco, the former panel supplier for the TF sports car, had also announced its involvement (Stadco 2006). The extent of Stadco's role is not clear, but it does suggest that the majority of the BIW production would take place in China rather than Longbridge.

SAIC and NAC also made some moves to divide MG Rover along human and physical asset lines. In buying the physical assets of MG Rover,

NAC did not initially pursue the human assets and a senior manager of the Chinese company described how the company had acquired at Longbridge the physical R&D capability to develop entire new models but without the manpower that had existed before. He confirmed that R&D would be centered on China, the role of a diminished Longbridge R&D being to "add a European flavour" and a sense of the MG heritage to future models. The continuing purpose of R&D in the UK was to "complement the heritage of the MG brand" and "understand the talents that are endemic in the UK industry" but without making a full commitment to UK R&D.

Indeed, NAC's strategy seemed to be to recreate the Longbridge operation, encompassing R&D and production capacity, at a new site in China. This would have left the original Longbridge plant as nothing more than the European branch of the operations centered on China. The result would have been a domestic operation in accordance with the full-function model of the automobile industry paradigm but with the same problems that MG Rover had suffered at Longbridge. This meant a lack of scale in production, an aging product range, and poor exploitation of international factor endowment differentiation by relinquishing access to MG Rover's R&D capability. However, a senior NAC executive subsequently confirmed that the plan had changed and that the company had decided to elevate Longbridge R&D capability so that it would be the source of all new model strategies. R&D in China would then focus on adaptations for the local market. To enhance continuity with the previous MG Rover R&D capability, NAC had rehired Mr Rob Oldaker to reprise his role as the MG Rover head of engineering, carrying the title of Global Product Director, and who would then lead the expanded R&D team in a partial recreation of the original MG Rover team.

In contrast, SAIC seem to have been much quicker in approximating to the automobile industry paradigm by acknowledging earlier the divisibility of functions by asset type and factor endowment. The company was interested in the human assets of MG Rover and these seemed to have locational advantages for the company. This was at variance to what had been expected of SAIC. After MG Rover collapsed in April, 2005, Holweg and Oliver (2005) speculated that SAIC would then be able to acquire the assets that it chose, namely the physical means of production. This is not, though, what happened.

Having already obtained the rights to various model and engine designs, the company eschewed a bidding war with NAC for the physical assets and instead rehired many of the human assets of R&D, i.e. MG Rover's development engineers, forming the Ricardo 2010 R&D unit under the guidance of Ricardo UK Ltd (DTI/Auto Industry 2007). A Ricardo media consultant provided an additional perspective for my research. He stated that Ricardo UK Ltd offered the concept of the 2010 unit and SAIC accepted because it provided access to the human capital behind the IPR formerly existing within MG Rover. He emphasized that IPR is not just the product design but also the systems and experience that lies behind it.

The media consultant asserted that SAIC's involvement in Ricardo 2010 was a low risk approach to securing technology transfer in the long term, contrasting this with consultancies that provide knowledge only for the term of the contract. In its initial stage the Ricardo 2010 unit represented a formal JV between Ricardo and SAIC, and it exploited the international factor endowment differential in R&D that MG Rover had originally offered SAIC in an IVJV. The 2010 unit recreated most of the MG Rover design team in order to preserve the organizational capital that resides within the human assets and so replicate the proposed IVJV. However, a problem for SAIC was that the organizational capital in fact resided with Ricardo UK Ltd, and as this is a dedicated R&D company serving a multitude of automobile manufacturers it meant that Ricardo UK Ltd could have gained *ex post* benefits at SAIC's expense. SAIC was dependent upon Ricardo 2010 for its R&D, but without the Ricardo UK Ltd in turn committing all its R&D capabilities to SAIC the relationship could not be defined as a true VJV. It was therefore important for SAIC to impose a greater degree of control over the relationship.

The IVJV strategy suggested by me could have been imposed on SAIC and Ricardo 2010 by demerging Ricardo 2010 from Ricardo UK Ltd. This would have established Ricardo 2010 as an independent party to the IVJV with SAIC. However, this does not solve the problem of a suitable governance structure for an IVJV that would formalize strategic control. In this case the solution came in the form of SAIC's complete ownership of Ricardo 2010. As suggested by Klein (2000), by taking ownership of the Ricardo 2010 R&D unit from Ricardo UK Ltd, SAIC secured access to all subsequent *ex post* benefits which might instead have accrued to Ricardo UK Ltd. It also gave SAIC control over the R&D strategy, though leaving the division with operational control. This seems to clearly reassert the M-Form of governance structure, yet I argue that ownership is not a necessary condition for the M-Form in an IVJV. Had MG Rover retained its status as an independent party to the JV, then the question of governance might have been answered differently. As it was, Ricardo 2010 could never have been independent because SAIC needed to retain links with Ricardo UK Ltd; SAIC therefore needed complete ownership to offset Ricardo UK Ltd's close operational connection with Ricardo 2010. An announcement made in 2007 stated that the advantages of ownership would still depend on the locational advantages of keeping the R&D unit in place so the Ricardo 2010 unit would be preserved within the physical boundary of Ricardo UK Ltd. According to a Ricardo UK Ltd press release (January, 2007): "the transferred business will continue both to be based at offices at the Ricardo Midlands Technical and have strong operational links with Ricardo UK Ltd."

Although the engineering team at Ricardo 2010 was continuing the work it had started at MG Rover, the engineers reported to the Ricardo media consultant that they had gained the one resource they had previously lacked: funding. The unit's task within SAIC was to conduct advanced engineering for new models, with a workforce of around 150, effectively the same size as the R&D department at Aston Martin. The engineering team includes not only some SAIC engineers from China but also representatives of Chinese supplier companies. By coincidence, the unit was operating in the same building that Ricardo UK Ltd had used in helping BMW finish the engineering on the new Mini, BMW having lost much of the R&D team when it sold its Rover Group assets.

With production and production engineering taking place exclusively in China, it was put to the media consultant that this replicated the IVJV suggested by my research; he agreed with this point of view. Just as Grossman and Hart (1986) asserted that a firm is one that has operational control of its assets, not necessarily ownership, the spirit of a VJV relationship between SAIC and Ricardo 2010 is embodied by maintaining Ricardo 2010 as operationally separate. This is somewhat reminiscent of the M-Form of organizational structure discussed in Chapter 4 and extends to Ricardo 2010 a measure of operational autonomy. Since the Ricardo 2010 facility exploits international factor endowment differentiation, as shown in Figure 7.4, the Ricardo 2010 R&D unit is operationally separate from SAIC, and therefore in an IVJV style relationship, even if the equity is owned by SAIC. This structure can also be applied to NAC, with Longbridge R&D taking the place of Ricardo 2010 and NAC replacing SAIC, although in addition the Longbridge plant would have its own limited final assembly capability from CKD kits.

With this structure SAIC was able to exploit the division between human and physical assets, as suggested by Dussauge *et al.* (2004), and an international division according to the factor endowment theory of Schott (2003). NAC, by purchasing mainly the physical assets and moving them to within close proximity of its main operations in China, initially recreated the full-function structure of MG Rover, but this time within the confines of the larger Chinese market. Although this would provide NAC with an entry into the local market, since the range of vehicles has been generally

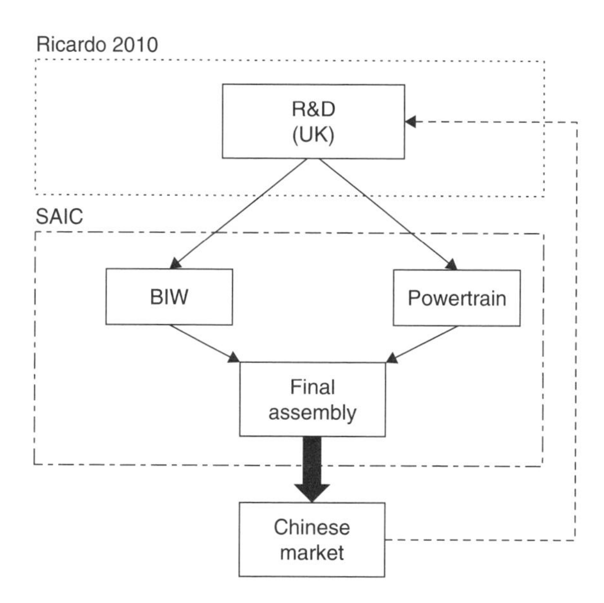

Figure 7.4 IVJV style relationship between Ricardo 2010 R&D and SAIC

considered to be uncompetitive in the global market the need to replace them was even more pressing than it was for MG Rover. Unlike SAIC, NAC had avoided the risk of hold-up when using an external R&D team by simply internalizing the capability. While SAIC can build on the IPR it has purchased using the team that generated it in the first place, it would have been mostly immaterial whether NAC attempted to update the product range that was purchased (as indicated to me by the senior manager) or design an entire new range; in either scenario NAC would have had to conduct R&D largely by itself, a function in which it lacked experience.

Although NAC had begun to follow SAIC in replicating the advantages of the IVJV structure by enhancing the role of Longbridge as a site for R&D, later events overtook the strategy. Under political pressure to construct a Chinese automobile group that is competitive within the global industry, SAIC and NAC (under the auspices of the parent group, Yuejin) entered into negotiations to explore collaboration. This has resulted in the absorption of NAC's automobile operations within the SAIC empire, coincidentally reinforcing the view that NAC's strategy of recreating the Longbridge operation in China was no more sustainable than MG Rover's original strategy. Furthermore, it shows that ownership of the physical assets of production did not seem to have conferred any particular advantage on NAC, which entered the market later than SAIC with its largely unchanged version of the Rover 75 and without any significant product

R&D programs. Only one process was unique in its physical assets, that of producing the advanced but ill-starred K-Series engine, and the physical asset specificity in this area was probably attractive to SAIC as it reportedly struggled to put its own version of the engine family into production. Again, though, it was not necessary for SAIC to own the assets since it could have contracted with NAC to supply the engines until its own engine family had been brought to the market.

The consolidation of SAIC and NAC has no material effect on the logic of the structure shown in Figure 7.4. The combination of the two companies does bring the usual benefits in scale as part of the overall trend in consolidation witnessed in the industry over the decades of the Budd Paradigm, but it does not affect the advantages that are derived from exploiting international factor endowment differentials. Furthermore, it would not weaken the advantages of maintaining semi-autonomous functions within the organizational structure, as could be contained by an M-Form of governance. In this way, the status of Ricardo 2010, now known as the SAIC UK Technical Centre, as an operationally separate R&D unit should remain unaltered and the organizational structure of Figure 7.4 remain intact.

7.4 Conclusion

In the previous chapter it was found that established automobile companies were mostly full-function MNEs that were either industry leaders within the automobile industry paradigm and producing above the MES_P, or were part of larger groups that were industry leaders. My research has found that they preserved their ownership advantages over their human assets by retaining the full-function structure originally formed within their home base. The governance structure for this strategy seemed to be provided by the U-Form, with certain adaptations necessary to include extensions into overseas markets. Although locational advantages and factor endowment differentiation were recognized, these were subservient to the need to exploit economies of scale at existing facilities.

This chapter has shown that MG Rover had the full-function structure prescribed by the automobile paradigm, but it was far from achieving the MES_P output. This has been attributed to its aging model line-up, leading to inadequate income being available for investment in future models. MG Rover was therefore exposed to all the cost disadvantages of being uncompetitive within the paradigm and only lasted as long as it had funds for its operational needs. While MG Rover had advantages in its high level of functional integration, as well as brand ownership and control of human assets in R&D, its problems stemmed largely from disadvantages in production output. The company had an assembly plant with output capacity that was within a range that many other companies accepted as being reasonably close to the MEPS, but to achieve the MES_P the company needed at least one more assembly plant, if not two, and the one it did have never reached full capacity by itself.

Even given its limited advantages, there was no evidence that the company understood its strengths or how globalization in its industry offered complementary solutions. Indeed, all representatives of MG Rover gave the impression that the company was operating with the same assets and opportunities as the established MNEs in the industry with output above the MES_P. Despite early statements concerning the search for a partner, it was only shortly before its financial collapse five years later that a suitable candidate, in the shape of SAIC, appears to have been identified. However, this was too late to save MG Rover.

The relationship between MG Rover and SAIC that was being considered conforms to the IVJV concept put forward in Chapter 6 as a means by which a firm operating inefficiently could best approximate to the automobile industry paradigm. The commercial logic of such a relationship remained even beyond the collapse of the British company. SAIC gained access to the IPR and related human assets of MG Rover through its involvement in, and subsequent ownership of, the UK based Ricardo 2010 R&D unit. This resulted in almost the same structure as the mooted IVJV, dividing the company functions into human assets (R&D in the UK) and physical assets (production in China), with additional benefits from factor endowment differentiation. Indeed, it was only the wider activities of Ricardo UK Ltd as an engineering consultancy that prevented the emergence of a VJV between two fully committed partners, and thus the necessity for SAIC to take control of Ricardo 2010 in order to safeguard its R&D knowledge base. The final structure is therefore in accordance with the IVJV model but secured by equity ownership.

NAC took a more circuitous route than SAIC. As the purchaser of the physical assets of MG Rover it seems to have initially attempted to recreate the British company in a Chinese setting. Access to a larger market would have brought greater production output and capacity utilization, but this was still far below the MES_P and there was no reason to expect the production capacity of the physical assets to have increased. NAC put in motion a plan to maintain Longbridge as an extension of the Chinese operations which would have ignored any possible advantages in international factor endowment differentiation. Subsequently, though, NAC took largely the same position as SAIC and elevated the role of the R&D capability in the UK.

From this it can be seen that the structure of the IVJV has been largely put into effect and an approximation made of the automobile industry paradigm, though not quite in the way originally envisaged by the theory. Rather than MG Rover remaining in existence as an independent part of the IVJV, the company collapsed, with the IVJV style of structure and benefits being largely effected by both SAIC and NAC. This has resulted in a full-function structure, but one that continues to recognize the benefits of endowment differentiation and the maintaining of a division between human and physical assets. This has ramifications for the global automobile industry, and for industry in general, which will be discussed further in the next and final chapter.

CHAPTER 8

The paradigm in the automobile industry of the future

So, where does all this leave us? The purpose of the automobile paradigm has been to provide a basic template for the companies that comprise the industry, yet if this hints at a static view then this would be to misconstrue the manner in which the industry has developed. Any neat prescription for a perfectly formed automobile company is bound to be beaten out of shape by the vicissitudes of the market place and the ambitions of the management. Like all paradigms, there is plenty of potential for development within the basic parameters. Elsewhere, we can see how the world's top physicists wrestle with the ramifications of Einstein's revelations while the rest of the world is happy to be pushed around by Newton's Laws of Motion. A paradigm only marks out the boundary of the battlefield, what happens within that boundary depends on the forces arrayed and the lie of the land.

The limits of the paradigm described in this book are set by the economic forces that structure the world's automobile industry. The paradigm then provides a systematic perspective on the organizational size and structure of companies that operate within the industry. In this book the resultant model is known as the automobile industry paradigm and it describes the conceptual limits of an automobile firm in terms of its most efficient size and structure.

The resultant automobile industry paradigm was constructed from two distinct perspectives: the first concerning economies of scale, based on the assumption of a common technology (see Chapters 2 and 3); the second concerning organizational structure, based on TCA (see Chapters 4 and 5). In separating these concepts I have been able to isolate factors relevant to firm size from those relevant to firm structure. It was then possible to estimate the minimum size and structure for a sustainable automobile manufacturer, known in this book as a prototypical automobile company.

Such a company should attain minimum efficient scale, known as MES_P for a prototypical company, along with a minimum organizational structure, which in this industry comprises the four main company functions (R&D, powertrain, BIW, and final assembly) of the full-function model.

If this were the end of the story then it would be a simple matter to describe the paradigm as a center of gravity for the industry, a point of convergence for all firms. The world, though, is far too unreliable for that. The ideal company form has been rejected as hopelessly out of reach, implying as it does unchanging production output of millions of identical vehicles. Instead, the prototypical company form is a compromise for a real world of variable demand and flexible manufacturing. In addition to this, there are many reasons why a company would wish to exceed the output prescribed by the prototypical model. In this sense, the prototypical company is one that has reached the first refuge in its growth strategy, the first point of safety, but in a capricious world it would need to grow still more in order to diversify away as many of the risks as was possible.

There is also plenty of scope for ambitious management teams to employ their overactive imaginations. Like the captain of an ocean liner, the fundamental task of senior management is to plot the appropriate course and send the company on its way. After the daily operational decisions of the lower managers, the senior management should only adjust the company's overall strategy when external events pose a serious threat. This, though, is to ask too much of senior managers that need to justify their existence. There are new products to be developed, production systems to be restructured, and shareholders to be impressed. The fact of the matter is that the product is still an all-steel body housing an internal combustion engine, while the production system is the same logical sequence of assembly operations that it has been for over a century. Attempts to reinvent either invariably end up as evolutions dressed up as revolutions. A true paradigm revolution would involve rather more than a car that has an electric motor that can come to the aid of the engine from time to time, or production refinements that shorten component order times. Developments such as these do show one thing, though: the automobile paradigm has plenty of life left in it.

8.1 Current state of the industry

Sheer size, measured in terms of production output and MES, has become a totem for the industry. In order to quantify the importance of scale in the automobile industry my research took two perspectives: the first, using survivor analysis, looked at the growth of market share for selected output ranges and the second calculated a prototypical MES_P from the estimates of MEPS. The first approach was rather inconclusive, since even over a 30-year period there were no indications that the largest output ranges were beginning to stabilize and it was not possible to determine how much further this output share would grow. At the same time, survivor analysis showed a second trend in the output ranges below 1 million units a year. Although the output share was falling, there were manufacturers within this trend that were holding on tenaciously, even expanding output, although at their rates of growth it could be a decade before they rose into the next output category. Survivor analysis therefore indicated two types of firm: the industry leaders with output above 2.5 million, and then a select number of smaller firms that seemed to survive at some minimum sustainable level.

The reason for the survivor analysis not being conclusive is that markets in emerging economies such as China and India are driving global automobile sales rapidly higher and so market shares of this growing total are still evolving. The alternative approach, calculating the MES_P from the MEPS for each function, was conclusive but resulted in an unexpectedly low output figure of around 600,000 units per annum. This referred to a company with a tightly grouped range of automobiles, each based on a steel unitary body construction but sharing technologies and production facilities where possible.

In parallel research, the necessary investment for R&D challenged the accepted view which is that it can cost up to £1 billion (US$1.9 billion) for a new model program. This research found that the continuing costs of R&D could be as low as £250 million (US$470 million at 2008 exchange rates) a year. The figure was derived from the latest DTI data from Subaru (the automobile brand for Fuji Heavy Industries), and even if the average figure over a four-year period is taken (2002–06) this only rises to £287.5 million (US$512.3 million at average exchange rates), with a peak figure of £313.3 million (US$512.2 million at 2003 exchange rates) in 2003/4. Furthermore, the 2006 DTI data for Proton in Malaysia, the only year available, showed R&D investment as low as £53.5 million (US$98.4 million) (DTI 2006), albeit with significant external assistance and a smaller model range.

The relatively low scale necessary for the MES_P indicates the fundamental condition of the industry. Since the production technology has been well established for the past few decades, at least since Toyota's TPS system refined manufacturing techniques in the 1970s, the industry can be assumed to have reached a mature stage in its life cycle. This would imply that the basic specification of the products should be fairly standardized, with profit levels having fallen and there being less product innovation.

In this situation, the strategy for these firms would be to reduce R&D and production costs to a minimum while defending market share through product differentiation, perhaps in market niches. This appears to be Subaru's position with a limited number of vehicle platforms underpinning a narrowly defined product range targeted at market niches (principally four-wheel drive). The strategy is reflected in the financial results, Subaru showing a net margin (net profit to revenues) of 2.1 percent for 2007. This is not as high as Toyota with 6.9 percent but is comparable with Mazda, showing a net margin of 2.3 percent, and Daihatsu with 2.1 percent (all figures from company reports). Without closely analyzing the different measures of financial returns it would seem, then, that operating close to the MES_P does not result in high profits but at least the business is generating an acceptable financial return.

The problem for a prototypical company is in maintaining output at the MES_P in an environment of regular model changes. This is shown by the financial data released for the fiscal year ending March, 2008, one that covered the beginning of a global economic slowdown centered in the United States. Combined with an aging model line, the net margin at Subaru was nearly halved to 1.2 percent for the year. The other companies benefited from the shift in the market towards more fuel-efficient passenger cars, Mazda's net margin rising to 2.6 percent and Daihatsu remaining steady on a 2.1 percent net margin. Although Toyota's recent move into full-size SUV sales in the United States exposed it to the worst aspect of the slump in that market, the company's diversified product range meant that a net margin of 6.5 percent represented just a slight downward adjustment.

Multiplying the structure, multiplying the problems

Although the MES_P represents a single automobile company operating at minimum cost, it is difficult to stabilize output at this level because of the fluctuations in demand. In practice, therefore, actual costs will rarely reach the minimum. The alternative strategy is to expand output in order to diversify the risk of output variation. The industry leaders have thereby reduced actual production costs by manufacturing at higher overall output levels from increasing numbers of plants. As the additional plants come on stream it is vital to raise demand in the market to support the new production capacity and ensure that all plants operate at their most efficient output level, known as the MEPS.

The problem with this expansionary strategy is that it can entice the company into deconstructing the unified full-function organizational structure that it had at the MES_P. This strategy is taken to its extreme when a manufacturer covers brands that have little in common with each other, in either production or R&D, each one of these brands representing a full-function structure by itself. For example, GM and Ford have long owned operations in Europe that are strategically and operationally separate from those in the United States. The European divisions do almost nothing to help the overall company to further exploit economies of scale. Although the individual plants may be operating at the level appropriate for the MEPS, the divisions they are allocated to may form semi-autonomous full-function structures that are fashioned like standalone companies. Far from helping the overall group to access economies of scale, they turn the corporate groups into multiple full-function companies or divisions. Each of these full-function divisions then needs to meet the requirements of the paradigm and achieve the MES_P by themselves in order to exploit fully the economies of scale available in the industry.

Ford has more recently been slimming itself down into something like a single full-function company, having divested itself of its British brands Jaguar, Land Rover, and Aston Martin, while at long last promoting its new small Fiesta as a true world car. Conversely, Tata of India, itself a relative newcomer, even in its home industry, is expanding with a multiple full-function structure encompassing the same Jaguar and Land Rover, along with its own Nano ultra-low price car and the mainstream Indica range. Each of these product families is distinct in its design, production, and even distribution to the extent that they are effectively separate automobile companies. For Tata to realize the economies of scale described by the automobile paradigm it must raise each to the status of a prototypical automobile company and achieve the MES_P at all of them. This implies a basic production level at each of 600,000 units per annum and a level of total output that would put it on a par with, for example, Suzuki of Japan, which is a single full-function automobile manufacturer. It should be noted that Suzuki is also by far the leading car brand in Tata's own backyard.

A multiple full-function structure confuses overall group output quantity with the ability of each division to mine the benefits offered by economies of scale. The logical process should instead be for an automobile manufacturer to first achieve the full-function organizational structure and the necessary production volume to achieve the same status as a prototypical manufacturer. From this position of the MES_P, the manufacturer should increase volume levels in order to access the quite separate benefits of stable production levels. As an industry leader with a broad range of variants derived from a limited number of base models, it is less likely to be inflicted with the punitive costs of production variability, unless a multiple demand failure should occur. This is statistically less likely given the breadth of the company's model range, but it is not impossible and can be catastrophic when it occurs. It is a predicament that has impacted most recently on Ford due to the restructuring of the United States domestic market, resulting in the closure of numerous plants as they are reconfigured for production of more fuel-efficient models. Nevertheless, even in its reduced state, Ford is a full-function company with output well above that necessary for MES_P, suggesting that it will be well placed to profit from the market recovery.

The full-function structure as the final defense

The restructuring of Ford is of crucial importance since the company, along with the other Detroit Big 3, GM and Chrysler, is being squeezed out of the SUV sector and forced to return to the domestic passenger car market, where lately the Japanese manufacturers have earned the greater part of their global profits. This could turn the local market into a pitch battle as the American manufacturers make a final defense of their home turf. The Detroit Big 3 could particularly exploit local advantage based on their empathy with US consumer tastes. This is most starkly demonstrated at motor shows where US domestic manufacturers exhibit flamboyant vehicles fashioned from swathes of sheet metal that evoke images of automotive land yachts cruising over wide open spaces. In contrast, Japanese manufacturers exhibit gadget-filled bundles of vehicular cuteness that seem more at home on the narrow streets of Tokyo. If the American manufacturers can exploit their home advantage to retake the lucrative passenger car market from the Japanese interlopers, then it might mark the end of the golden era for Japanese manufacturers.

Unlike the American firms, Subaru does not have the resources to consider downsizing to the home market, and its response to the difficulties of demand fluctuation has been to refocus its resources on its core product line. It has formed an alliance with Toyota, which has in turn taken a 17.5 percent equity holding in Subaru in order to secure the relationship. One area of Subaru's business where it was a long way from achieving economy of scale benefits was in the manufacturing of minivehicles, a range of diminutive models specific to Japan and marketed in accordance with government incentives. From 2010, Subaru will relinquish responsibility for these models to another Toyota affiliate, Daihatsu, which will then supply some of its own models under the Subaru brand. The minivehicle market in Japan is a fascinating Lilliputian world of hatchbacks, sports cars, and

SUVs that, to all intents and purposes, represents a parallel market of pygmy vehicles. The parameters of the automobile paradigm still apply, though, and throughout 2007 only Suzuki and Daihatsu came close to the 600,000 unit output prescribed by MES_P. During the same period, Subaru sold just 140,000 units.

In abandoning this market sector and concentrating on a core model range, Subaru is demonstrating how far other manufacturers have strayed from the pursuit of minimum cost. In effect, Subaru is planning to derive its entire model range using one platform as a common denominator. The model ranges will span from the compact Impreza to the mid-size Legacy and seven-seat Exiga multipurpose vehicle (MPV), along with a number of other variants. Such extreme consolidation around a common platform is probably going to be a feature of the wider industry as companies relearn the benefits of economies of scale; VW, for example, pledging to have 95 percent of its models on just three common platforms. In its alliance with Toyota, Subaru will take this further by also spinning off a sports car variant of which it and Toyota will have their own versions on sale from 2011. Apart from providing Subaru with greater production capacity utilization, this will allow Toyota to market a sports car again, after a gap of over four years. Perhaps most interesting is that the project appears to signal Toyota's tacit agreement that Subaru is more cost effective in model development. The fact that development costs for a low volume model must be minimized is irrelevant: developments costs should be minimized for all models, regardless of production rates.

Pooling resources and the international vertical joint venture

Subaru's relationship with Toyota is hardly a new feature of the industry. JVs have always been used as a vital crutch to manufacturers that are weak in critical areas. As was shown in Chapter 6, a scale alliance provides a route towards achieving an MEPS by filling output capacity in combination with a partner. Most clearly observable in production, there are many examples where it has formed the basis of a stable alliance. GM shares production facilities with Toyota in the United States at NUMMI, and with Suzuki at CAMI in Canada. Nissan and Chrysler are taking the scale alliance concept a stage further with a reciprocal deal that will have Chrysler supplying pickups manufactured in Mexico to Nissan, while Nissan will supply Chrysler with a new compact car. The pickups assigned to the Nissan brand will combine Chrysler platforms with Nissan styling. In return, Nissan will manufacturer a new compact car in Japan from 2010, a version of which will be styled by Chrysler for marketing under its own brand. What is so interesting about this deal is that it provides both partners with the benefits of economies of scale in manufacturing, but it maintains product differentiation by separating their vehicle design capabilities.

Where companies do team up in sharing knowledge the relationship is known as a link alliance. These usually provide shared access to the technical capabilities of vehicle design but, as I have demonstrated, there are risks of *ex post* opportunism when one partner finds that it has gained all that it needs and pulls out. Consequently, these alliances tend to be short lived, as one partner sees no long-term future in it. The industry leaders tend to engage in them on a project-by-project basis in order to spread the risk of such opportunism occurring. The fact that Chrysler and Nissan have been careful to avoid mixing their R&D contributions in their alliance indicates that they are evading the same dangers.

Nissan and Chrysler are both manufacturers looking for a solution to the short-term vicissitudes of demand as the United States market restructures around more fuel-efficient vehicles. In terms of the automobile paradigm they are both well placed to enjoy a sustainable future once they have stabilized. The two companies have production capacity well in excess of the 600,000 unit per annum threshold prescribed by the automobile paradigm, and they both contain the four main activities laid out in the full-function organizational structure model. Chrysler's financial situation deteriorated gravely in 2008 as it struggled to keep up with the shift in market demand from SUVs to passenger cars. The short-term solution is a massive injection of cash so that it can realign its model mix, but as a corporation it also needs to restructure drastically. This should involve consolidation of its model ranges around a small number of shared platforms, as well as consolidation of its production facilities. Nevertheless, Chrysler has an advantage of scale and capability inside the most important market in the world that few can rival.

This is not the case for automobile manufacturers that are operating well below the MES_P since their lack of capacity means that their first task must be to raise output to the threshold level. I have shown that MG Rover was in this sub-600,000 unit per annum output range, and I have suggested that an IVJV would have created a beneficial division in the full-function structure. This would have avoided the kind of *ex post* opportunism seen in link alliances by returning all the benefits of a new development to the partner responsible for the function. Opportunism would therefore be expressed as innovation and of greatest benefit to the firm that developed it. We have already seen some of this arrangement being observable in the separation of vehicle design in the Chrysler–Nissan partnership. For an

IVJV, additional benefits can also come from international separation as the contrasting factor endowments of different regions can be exploited to mutual benefit. The IVJV is therefore not simply a device that could have salvaged MG Rover's R&D capability but also a strategy for raising an assembly partner to a competitive position by providing immediate access to current product designs.

The weakness with the IVJV organizational structure was found in formulating an appropriate governance structure. This research had found little support for the M-Form of multidivisional control amongst the industry leaders, concluding that modern production systems favor the more centralized U-Form of governance structure that binds operational and strategic concerns together. The problem here is that the IVJV is instead founded upon a clear operational separation of functions. In purchasing the assets of MG Rover, the Chinese firms SAIC and NAC preempted the possibility of forming an IVJV with an independent MG Rover. However, SAIC, and to a lesser extent NAC, partially replicated the IVJV structure by internalizing the R&D function of MG Rover while granting it a degree of operational autonomy. This resurrects the M-Form with the specific purpose of exploiting the local factor endowments, namely the engineering talent available in the UK. In accordance with the M-Form of governance structure, strategic control is separated and resides with SAIC. The difficulty for this research is that this does not indicate where strategic control would have resided in a true IVJV between independent partners. This is particularly relevant for automobile companies where R&D and strategy are closely tied. A possible answer may be found in the kind of loose conglomerations exhibited by Japanese *keiretsu* groups where cooperative relationships are secured by minority equity cross-holdings. A governance structure appropriate to an IVJV is therefore an area where a continuation of this research could usefully be focused, Sako (2006) having already made a valuable contribution in her research on Toyota's company boundaries.

Diversity in an expanding market

Assuming that the IVJV structure offers a route to sustainability for many disadvantaged automobile manufacturers, then, rather than the convergence towards a half-dozen global corporations that has so often been predicted, we should see a healthy diversity of independent companies. The Chinese industry could well be key here since it is located within a rapidly expanding local market and a large labor pool. What the domestic industry lacks is the technical capability to compete with the global giants on the same level. During a period of relative safety under the protection of the Chinese government, these companies are learning from the industry in the rest of the world. Once the Chinese industry has reached the international standard, then the size of the domestic market should offer enough potential sales to sustain its operations. Of course, the counter trend amongst manufacturers is to raise output over wider product ranges in order to diversify the risk of variable demand. Yet, for as long as the global market continues to expand, there will be opportunities for smaller manufacturers to enter the industry and find stable demand for their products.

Output and forecast volumes published by *Automotive News* (2008) suggest that total global light vehicle production, the vast majority of which are comprised of automobiles, could grow from 55.2 million units in 2007 to 65.4 million units by 2012. Figure 8.1 illustrates this, differentiating the actual data for 2004 until 2007, and the forecast data for 2008 to 2012. Much of this growth will occur in China, where output should expand from around 9 million units in 2008 to nearly 14 million units in 2012. According to the World Business Council for Sustainable Development (WBCSD)(2004) the Chinese passenger car market will rival that of the United States by 2030, although forecasts concerning the Chinese market often underestimate the growth in demand. Suffice to say, there is enough room in the market for competitive newcomers to stake out their territory.

Were the status quo described by the automobile paradigm to continue indefinitely, then the dominant trend would be towards corporate

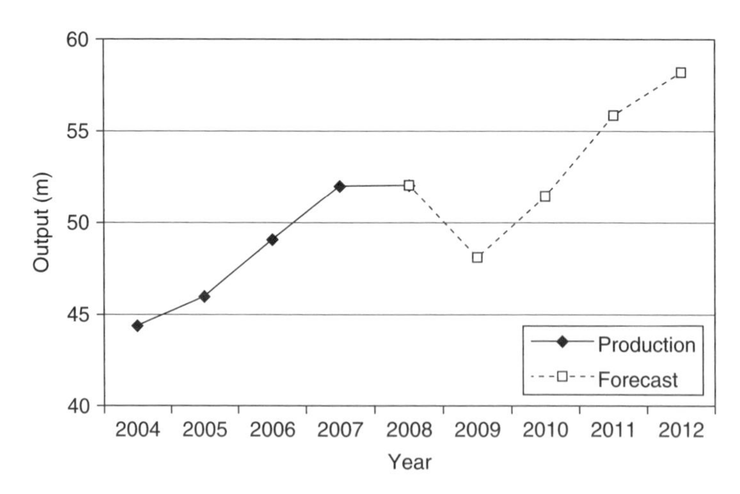

Figure 8.1 Light vehicle production output and forecast, 2004–12

Source: *Automotive News* (2008).

convergence simply because the only long-term security is in higher output over wider product ranges. This point, though, is an academic one since in all probability the oil reserves on which automobiles are so dependent will have been enfeebled beyond the point of utility sometime after the middle of this century. The industry leaders are well aware of this and are directing a large proportion of their R&D budgets towards untested technologies. Yet it is the very uncertainty of this condition that will force a paradigm revolution on the industry, with no respect for the established positions of the global giants.

8.2 Reaching beyond the automobile industry paradigm

The observation that the industry leaders are making greater commitments to R&D was one of the more interesting results of this research. It was expected that R&D costs per model would have been lower for larger firms due to an increased number of model variants over a higher total output volume. Instead, the industry leaders were found to be spending more per model than smaller manufacturers. Figure 8.2 shows how Toyota's 330N platform for the Corolla came to be shared across six models by 2008, but with output ranging from 1.2 million units for the Corolla itself to around 24,000 units for the Blade. Although this appears to represent the advantages of using a common platform in order to diversify the product range,

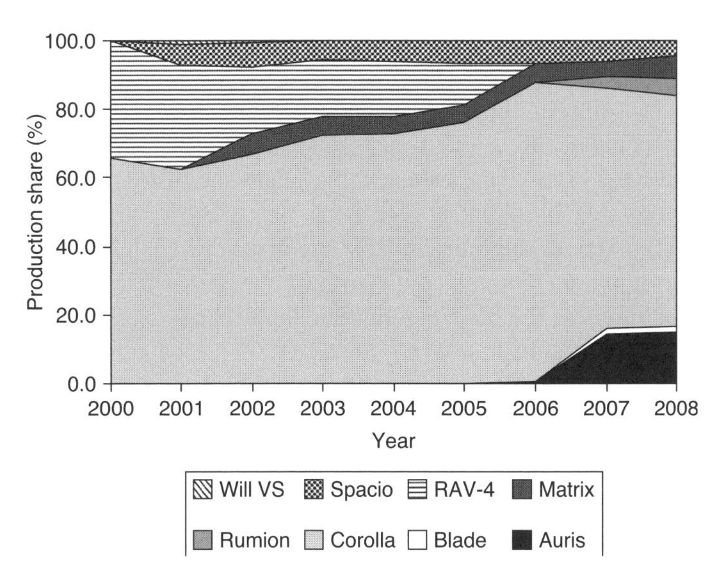

Figure 8.2 Proliferation of Toyota Corolla platform 330N

Source: IHS Global Insight.

it also demonstrates the temptation for a company to create a new model simply because it has the resources to do so. For niche models the marginal benefits may be inconsequential and, as previously noted, for its new sports car Toyota has turned to Subaru for the design work as well as production.

The very largest automobile manufacturers have the resources to engage in activities outside of the paradigm. In R&D, for example, Honda has committed itself to a range of products as diverse as racing-car engines, jet engine design, robotics, and fuel cell research. In the course of my research, Honda company representatives attempted to justify these programs, even with a degree of validity, but it should be emphasized that these new technologies cannot be included in the current automobile paradigm.

Concept cars that are developed for motor show exhibitions have some claim to being of greater relevance to future production models. Yet again, though, there is some doubt about the cost effectiveness of such programs. At the 2007 Tokyo Motor Show, Toyota exhibited eight concept vehicles, one of which even demonstrated an ability to adapt itself to the emotional state of the driver, yet none of them were expected to reach production in their complete form. In contrast, Subaru exhibited just two concept models: the G4e electric vehicle proposal, based on an existing production model, and the Exiga seven-seat concept car, which was released as a production model eight months later.

The global automobile giants do not restrict their corporate diversions to even these vaguely automotive related adventures: they may also step outside the automobile paradigm and integrate with external activities. Under normal conditions these are activities that should be contracted for in the open market, but where the market is poorly developed then the automobile manufacturer can secure access to supplies by integrating with external services. In the early years of the industry this is how the full-function structure came to evolve. Consequently, GM, Ford, and Chrysler all had their own components divisions to ensure reliable supplies of ostensibly generic parts. Reliability of market demand could similarly be enhanced by internalizing finance providers in the days before high street banks were ready to offer loans to first-time car buyers. The General Motors Acceptance Corporation (GMAC) was formed in 1919 to extend credit facilities to its customers and thereby played a vital role in promoting sales of the company's models.

However, once the industrial and market landscapes were fully developed, there was little need for automobile companies to retain operations that lay outside the boundary of the full-function model. Yet the rationale persisted that hyperextended vertical integration permitted access to lower cost supplies as well as additional income flows. These, though, are entirely false premises as inspection of the value chain shows.

The value chain is a graphic representation of the flow of materials, parts, and services that culminate in the finished product. Each input has an additive impact on the value of the product with cost implications for the final selling price. Although this is a fair illustration of the production-transformation process it also presents two tempting strategy distractions for senior managers. The first is that any additional activity will add value which can be recovered in the higher price of the product. This is analogous to classical economics where inputs define the final price of a product, a view that ignores the crucial role played by market demand. This can lead a company into installing new features in a product which are then rejected by customers, resulting in higher production costs without any increase in perceived value. Since the customers would resist any price increase to recover the additional cost, the final price is effectively discounted. Fortunately, within the automobile paradigm, such minor experiments are also a healthy part of the heuristic development of new products.

The more serious distraction is one that takes the automobile manufacturer into entirely different sections of the value chain. This includes the original extensions into component production and financial provision seen in the formative years of the industry, but it is a strategy that has continued into the modern era. The belief that such a strategy provides access to lower costs can be readily dismissed: TCA of the automotive industry has shown that only the four main functions of the full-function model result in cost advantages when they are integrated. There is no particular reason why GMAC, for example, should be more cost effective in the provision of financial services than any other financial organization in the market. Granted, as GM's captive provider, GMAC enjoyed unparalleled access to GM dealerships, but any financial benefits would be attributed to monopoly power rather than any economies of vertical integration.

It is undeniable that hyperextension into activities outside the automobile paradigm creates new revenue streams, but this would be the expected outcome of any normal investment strategy. The fact that the external organization is also one that trades with the automobile firm on an operational basis is irrelevant. As an investment strategy it helps to diversify financial risk so that a fall in revenues in the core automobile sections can be offset by revenues in, for example, the finance arm. This has the effect of maintaining group revenues, so stabilizing the creditworthiness and share price of the firm as a whole. However, if the cross-subsidization becomes a permanent fixture then the finance arm will suffer a cash drain that it could have used for investing in its own successful financial activities. Furthermore, propping up the automotive side of the business will defer necessary remedial action until a crisis point. This is particularly pertinent in a general economic downturn when both branches of the business may suffer at the same time. This is precisely what has happened at GM and GMAC during the economic problems centered on the United States in 2008, though fortunately GM had sold a majority stake in GMAC in 2006.

Paradigm revolution for the automobile industry

Research that is being conducted on new technologies represents exploration of the next paradigm for the industry. As far as vehicle motive power is concerned, it is well known that oil is both a limited resource and a cause of environmentally harmful exhaust emissions. Hybrid technology, which combines an internal combustion engine with an electric generator, is often cited as an interim step towards an alternative technology. However, since it retains an orthodox internal combustion engine, it is adding significant cost to the production process. A joint study by UBS and Ricardo Consulting found that a hybrid powertrain, as fitted to a Lexus RX400h, was twice the cost of an equivalent fully compliant clean diesel engine and up to five times the cost of an equivalent petrol engine (UBS/Ricardo 2007). Toyota has conceded that despite successive cost reduction of 50 percent on its Prius hybrid model, it is only the third generation version, to be available after 2010, that will earn profits comparable to a mainstream model. In effect, hybrid technology adds another costly function to the full-function model, namely that of electric powertrain, and so expands the scope of the current automobile paradigm but does not supersede it.

A new paradigm for the industry should involve a distinctly new technology that renders the current paradigm obsolete and would thereby herald a paradigm revolution. Fuel cell technology is often promoted as this technology because it not only depends on a new motive power, using electricity generated from a catalytic reaction between hydrogen and oxygen, but the system also has implications for the entire production system. Whereas a traditional engine is a bulky device that can only be positioned in a limited number of ways within a vehicle, generally at the front or the back, a fuel cell system can be positioned with more flexibility. The 2008 Honda FCX Clarity fuel cell vehicle, for instance, holds its fuel cell stack in the transmission tunnel between the front seats. As a consequence, the ancillary components, such as batteries and fuel storage, can be distributed throughout the vehicle platform.

GM designed the "AUTOnomy" concept as a chassis that contained all the hardware of the fuel cell system, allowing the body design to be entirely separate (Nash 2002). Like the craft style of automobile production in the early years of the industry, this concept means that the vehicle body is not the main load-bearing structure, so there is greater flexibility in body design and use of materials. This in turn makes the steel unitary body process of the Budd Paradigm redundant, taking with it the associated economies of scale in the industry. The AUTOnomy concept opens the possibility of a resurgence amongst independent coach builders, able to supply custom-made bodies in small numbers while automobile companies concentrate on mass production of the chassis and powertrain.

This would be feasible if it transpires that the economies of scale for fuel cell manufacturing are very high, particularly in view of the capital intensive nature of production, or that consumers consider them to be a generic product differentiated only on price. Automobile manufacturers might then find that they need to restructure their hitherto full-function integrated structures. It is conceivable that some manufacturers might even specialize in the production of fuel cells and relinquish their vehicle body production facilities. If this were the case, and GM's AUTOnomy concept became the industry standard, then automobile firms could manufacture generic vehicle chassis in a central location and ship them to small, final assembly plants that served their local markets in the style of micro-factory retailing (Nieuwenhuis and Wells 2003). This, of course, would only realize economies of scale if body production was suitably adapted to a new MEPS, suggesting the use of materials other than steel and the unitary body.

While all this seems to herald a bright new future for personal transport, fuel cell technology is still at the experimental stage and it is not known when, or even whether, it will reach mass production. Honda released a version of its FCX Clarity in mid-2008, but this was a strictly limited production run of 200 units made available for lease to selected customers in California. We can speculate on the reasons why this area was chosen, such as the receptive self-styled hippy culture and the highly successful local economy, but it is also likely that the climate will be kinder to the delicate fuel cells. The payments have been set at US$600 (£320) per month for 36 months, or US$21,600 (£11,550) in total. The real cost of the vehicle is likely to be nearer US$1 million (£530,000) at least, and the cost of the raw materials used in the fuel cell means that they are unlikely to ever be cheap. Furthermore, there is no significant hydrogen fuel infrastructure in place and, though Fuel Cells 2000 (2007) lists around 30 hydrogen fuel stations in California, only a small proportion are open to general use. As a consequence, customers of the FCX Clarity are required to have their own home refueling station. Such a commitment may mean that the first fuel cell applications will be better suited to fleets of commercial vehicles with a centralized fuel supply. Alternatively, fuel cell vehicles could manufacture their own hydrogen from other fuel sources carried on board, but this would add yet more cost and complexity to the vehicle.

Since the future of fuel cell technology is unknown there is no way of telling if the research strategies of the automobile industry leaders are the appropriate ones. The uncertainty surrounding the possible paradigm revolution means that automobile firms that are operating at or near the MES_P in the current paradigm will not necessarily suffer a disadvantage in the new paradigm. Indeed, should the MEPS for any given function be significantly reduced, particularly for body production, then these manufacturers may be as well placed as any larger manufacturer. Companies such as Mitsubishi Motors, which is investing heavily in electric vehicles, could thereby suddenly find itself at the forefront of the industry.

There may still be a place for hydrogen if it is consumed within an internal combustion engine in place of oil-based fuel. Although BMW developed the first hydrogen fuelled car, the 750hL, in 2000 (BMW 2007), using a conventional engine, Mazda's rotary engine technology is more suited to the consumption of hydrogen fuel (Yamaguchi 2003). A switch in the industry to production of rotary engines would shift the present paradigm, but not enough to mark a leap to an entirely new paradigm.

The research that is continuing in conventional technology indicates that there is plenty of life left in the oil-fuelled automobile paradigm. The internal combustion engine is still only 30 percent efficient at best and research has been directed at releasing more of this potential. Proposed engine designs, such as Ricardo's 2/4sight which can switch between two- and four-stroke modes, offer impressive increases in fuel efficiency. At the same time, lighter vehicle bodies, thanks to the use of aluminum, amount to an evolution in production technology that signals a subtle shift in the paradigm as opposed to a full revolution. Overall, then, the pace of research within the present paradigm suggests that the revolution may be a long time coming.

8.3 Paradigmatic views of other industries

The automobile industry paradigm, like all paradigms, is founded on a defining feature. In this case it is the production technology related to the steel-welded unitary body that forms both the shape and the structure of the automobile. In other industries there will also be defining features that then define the basic processes and the economic forces that mold the industry. In the aircraft industry it is again the assembly process that dictates the size of the companies involved, it being advantageous to concentrate production in a small number of massive plants. This paradigmatic approach is not restricted to manufacturing processes; indeed it can be used to clarify our perspective on almost any activity from health care to shopping. For example, the retail industry is centered upon consumer behavior, and as much as retailers may wish to reduce costs by minimizing, or even eliminating, inventory, the desire of shoppers to "try before they buy" will always require the provision of shop displays at the local level.

The benefit of a paradigmatic perspective is that it permits a highly structured representation of an industry, whatever it may be, in order to identify economies. This, then, offers promise to the global divisibility of structures and the potential for an IVJV. However, the IVJV strategy is only suitable in exceptional circumstances. For a suboptimal firm to exist at all suggests some form of underlying support, perhaps through government protection, company cross-subsidies or short-term and dead-ended consumption of internal company resources. When two companies are suboptimal to the relevant industry paradigm in complementary ways, then there is the potential for them to form an alliance that allows them to take sole responsibility for their strengths, while relinquishing their weaknesses to their stronger partner. Furthermore, they should be located in regions with contrasting factor endowments so that they can exploit the comparative advantages. The final result, in order to replicate to their industry paradigm, should be a complete industry-standard set of functions (i.e. full-function) that is divided around the globe but with overall output that is competitively close to the MES for the industry.

The shipbuilding industry might have potential in this way. At one time shipbuilding was a major industry in the UK, a country that retains many of the skills relevant to naval architecture. For example, Harland and Wolff in Belfast is famous for designing and building ships for the historic White Star Line, most notably the *Titanic*. While the company retains a ship repair facility, it no longer builds ships, although it continues to have a ship design capability (Harland & Wolff 2007) that would permit it to link with a shipyard elsewhere in the world. To illustrate this, Figure 8.3 shows an example of an IVJV between Harland & Wolff and an overseas shipyard in an emerging economy to exploit low costs in production. As in the automobile IVJV, the market intelligence would return to the company that initiated the product, Harland & Wolff. Interdependency would be created by Harland and Wolff's need to have access to a shipyard, while the shipbuilder would need to have a supply of ship designs. The interdependency obviates the necessity for rigid, exclusive contracts and it is likely that both could pursue their own strategies and therefore their own separate

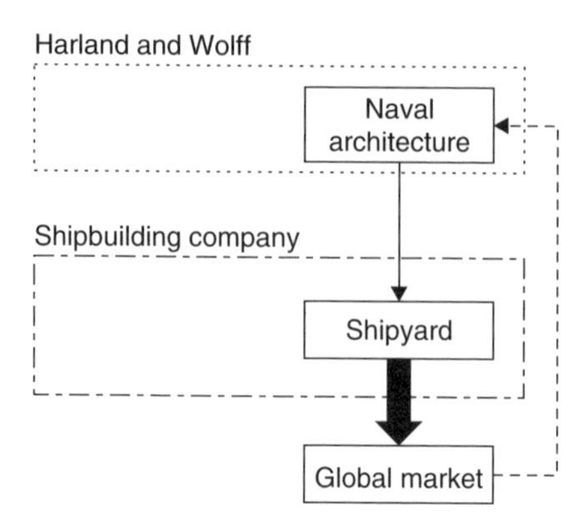

Figure 8.3 Suggested IVJV between Harland & Wolff and a shipbuilding company

governance structures. A similar relationship seems to be evolving between Carnival Corporate Shipbuilding, designers of the Cunard cruise liners, and the Italian shipyard, Fincantieri.

For a complete IVJV, Fincantieri would have to relinquish all cruise line design to Carnival Corporate Shipbuilding. Such a dramatic strategic shift is often politically unpalatable in any country since it means that the nation must lose a capability it has long nurtured and become dependent on another country. For this reason, there is always robust political resistance to joint ventures in the armaments industry, as it is important to balance the cost advantages of foreign production with the need to maintain a strategic manufacturing facility in the home market. Consequently, countries strive to be self-sufficient in all aspects of the defense industry. The tragic irony of this strategy is that it misses one of the fundamental advantages of the IVJV alliances: they are founded on a cross-border dependency that could serve to assuage international tensions, thereby reducing the needs for strong defense industries in the first place.

Implications of the paradigm for national policy

My research has found that globalization is not an ultimate state but a continuing process of cross-pollination between distinct geographic regions based on their factor endowment differentiation. A narrow view of factor endowment differentiation would suggest that globalization will eventually lead to a homogeneous global industry as countries converge on a common level of industrialization. A broader view, though, as espoused by Schott

(2003), shows that factor endowment differences are generally stable and so globalization is a dynamic process as regions continue to develop their specific advantages. For example, China with its massive population will be relatively wealthy in terms of its labor resources for as long as it maintains its demographic advantage.

Leading companies in the automobile industry have tended to ignore the opportunities offered by globalization because they find greater advantage in their centrally controlled internal markets. This has resulted in industry leaders taking a strategic view of opportunities, locating plants as part of an overall global plan rather than to exploit specific local advantages. It is something of an irony that Japanese automobile firms have selected the UK for production plants when the UK's comparative advantage is in engineering rather than labor intensive assembly work. Since the rationale behind the plants is strategic, they may be moved to other locations should the strategy require it. Ford's conversion of the Dagenham final assembly plant into a center of excellence for diesel engines is an example of this; as a consequence, all Ford branded cars are now imported into the UK, principally from assembly plants elsewhere in Europe.

Dunning (1979) suggested that factor endowments can be developed further by investment in facilities and workforce training. Given that the definition of any forthcoming paradigm revolution in the automobile industry is clouded with uncertainty it could be argued that research outside of the current paradigm should be the preserve of national governments, not automobile companies. This would then create a bedrock of capability that manufacturers could draw on by cooperating in specific programs or recruiting staff that have experience of the latest developments. Government research would thereby provide knock-on benefits for the national economy in a similar manner to defense programs or space research.

There is nothing new in this suggestion, and governments have at times become involved in company operations at times of crisis, but with only sporadic success rates. Indeed, MG Rover and its predecessors were subjected to government involvement in more ways than perhaps any other company in the industry. Part of the reason for MG Rover's final collapse was that the UK government could not, and would not, rescue it again. This is not, though, what the development of factor endowment should entail. If the skills that were prevalent within MG Rover have been redeployed elsewhere in the economy to greater net benefit to the country then the government was wise not to intercede in the company's fate. Government policy should instead be directed at ensuring these fundamental capabilities are progressively enhanced.

Of course, there is no certainty that a government is more effective at identifying which factor endowments to support than it is at identifying which companies to rescue, and my research has found little gratitude in the industry for government involvement so far. For example, one senior executive of MG Rover registered his disappointment with the quality of graduates emerging from British universities. National pride can also cause obstructions, such as the Chinese government promoting the development of R&D capability in the domestic automobile industry, even though the nation's comparative advantage lies in production due to the greater labor resources. If the myriad of new Chinese car manufacturers are found to lack the absolute level of R&D capabilities with which to compete against the global industry leaders, then they will fail to reach the MES_P and will require continuing government support. If they are not given artificial support then they will be simply ejected from the industry. Instead, the Chinese government should enhance the nation's underlying production capabilities so that it retains its current production advantages. Perhaps controversially, this implies repealing the "one child" family planning policy and allowing the population to expand unfettered. The IVJV concept would then provide the linkage between the vast local labor resources and the technical R&D capabilities elsewhere in the world. The mechanism for achieving this is free trade, allowing companies to seek out resources in an open market. The role of the governments, then, is to sponsor national factor endowments within a secure legal framework so that companies can structure themselves around the global opportunities in the most efficient manner.

Suggestions for subsequent research

One of the most surprising discoveries of my research was the relatively low quantities of output required to exploit the economies of scale achievable at the MES_P. This was based on survivor analysis and a calculation from the MEPS for each function. I have often cited Subaru, but recently Chrysler and Daimler have been reintroduced to lower levels of output in the wake of their demerger. At least in the case of Daimler, this appears to have resulted in sharply lower costs. Below the minima prescribed by the MES_P there are, of course, numerous Chinese manufacturers all attempting to reach a state of sustainability while they are still under the protective wing of the Chinese authorities. Once the Chinese market is fully liberalized then the domestic manufacturers will be exposed to the global industry. There is, therefore, a sense of urgency for them in achieving the MES_P before the government support falls away. For this reason it is important for other studies to corroborate the measure of the MES_P that I have ascertained.

In this book I have attempted to provide a cogent argument for IVJVs, but there has been, so far, little empirical evidence of them occurring in their pure form. There are, though, plenty of signs of the IVJV structure in other alliances – such as Rootes Group and Iran Khodro, Nissan and Chrysler, and so on – but I have not found any such relationships in their definitive form. This is partly because they are only possible in exceptional circumstances, but also because there is political resistance to cross-border interdependencies, particularly in key industries such as automobile manufacturing.

Another reason why the IVJV has been slow in emerging is related to the particular problem I have had in formulating an appropriate governance structure for the IVJV. In discussing various governance structures (see Chapter 4) my research has found little empirical support amongst the established global giants for the much-vaunted multidivisional M-Form, yet there may be some possibility that it would be suited to the IVJV in recognition of the clear separation of operational functions. This then leaves the problem of where to allocate the strategic control. Fortunately, this may not be the critical problem that it at first appears since automobile markets are similar enough to accept common vehicle designs. Evidence for this can be seen in the way that SAIC and NAC put the Rover 75/MG ZT into production in China with few changes, apart from a slight increase in rear passenger space. Conversely, conducting R&D within the target market may not necessarily result in increased sales, the Honda Ridgeline truck being designed in the US but capturing relatively few sales. This suggests that it may be possible to allocate R&D strategy to the R&D partner since they are as well placed as any to judge the needs of the market.

8.4 Closing remarks

The paradox of a paradigm is that although it represents a fixed point of reference within the field of knowledge, at the same time the intensity of problem solving that goes on inside the paradigm means that it is far from being an oasis of calm. The automobile paradigm is no exception. On the one hand we have a product that has gone about its business in much the same way for over a hundred years, along with a manufacturing system that has not fundamentally changed in around 80 years. Yet within these parameters the implications are still being worked through at a tremendous pace. Even now, no manufacturer can claim to embody perfectly all the principles set out by the paradigm. The industry is awash with the unremitting flow of innovations, from super-clean, super-efficient engine technology to the leanest of production methods. It is in the nature of industrial innovations that they are promoted as revolutionary leaps into the future even though, with a little sober reflection, they can be more properly seen as evolutionary solutions to perennial problems. In the same vein, globalization has been depicted as a tide that is sweeping everything before it, when the prosaic truth is that it is progressively raising industry to a higher level. This is not to say that the next paradigm is not waiting over the horizon, but when it does appear it will be in a form that almost no one could have anticipated. In the meantime, the industry's future lies in the present.

EPILOG 2009: THE BATTLE FOR THE UNITED STATES

It is the recurring tragedy of the world's car industry that companies must bet the farm at the start of every new product life cycle. It is because the capricious consumer demands a new model at regular intervals that the manufacturers are doomed to replacing even their most popular vehicles with new offerings of unknown potential. Since the automobile industry paradigm shows that there is a monstrous production system to feed, the failure of a core model can put the company's future in peril even after it has successfully defended its place in the industry over many decades.

Although the automobile paradigm prescribes a reasonably achievable level of output for exploiting the available economies of scale, this is contingent upon it being reached with production consistently at full capacity. Firms operating close to the minimum standards of the paradigm must stabilize demand by releasing highly differentiated models into market niches. Those that cannot push output to this level must cover their higher costs through premium pricing. One small manufacturer, Porsche, has even managed to use its premium profits to practice a strategy that I call "cuckoo-nesting." This is when a small, highly profitable firm is able to take over a larger one in order to stake a claim on its choicest resources. In effect, Porsche is using VW as the unwitting host for much of its model development and production.

At the other extreme, the sheer size of the industry-leading firms gives them the additional benefit of spreading the risk of fluctuating demand over a wide range of model variants. This effectively insures the industry leaders against the effects of individual product life cycles and isolated instances of model failures. What this strategy cannot withstand is a multiple product failure. To the horror of the world's automobile industry, this is precisely what happened in 2008 and 2009, as the economic shocks rattled through the markets like earthquakes. In combination, these convulsions have shifted the tectonic plates of the world vehicle markets and left the manufacturers desperately trying to restructure around the new demand characteristics. We can have faith that the economic recovery will eventually come, or else this truly is capitalism's Armageddon, but the question is which survivors will emerge from the rubble.

The automobile industry is all about mass production feeding mass market demand. In the pre-earthquake era, although some markets had already slowed to mature levels of modest growth, emerging markets were still surging ahead. The generally held view was that success lay in expansion so that, even if the developed markets stagnated, salvation would be supplied by the new developing markets. The first seismic jolt to upset this view was the financial cataclysm that came out of the United States and rumbled through the world markets at the beginning of 2008. While this unsettled many car manufacturers, it should have been survivable. The trouble was the shocks just kept coming. The collapse of Lehman Brothers plunged the credit markets into chaos just as recessionary pressures crushed the last vestiges of consumer confidence. Taking cover from the successive onslaughts, consumers shrank from the market and the car manufacturers were brought to a shuddering halt.

We would expect this to kill off some of the more bloated corporations as market shifts leave them marooned with model ranges suddenly no one wants. Most staggering of all, though, has been the way in which this disaster has swept the ground from beneath the Japanese manufacturers. Amongst their number, Toyota is preeminent and in the early stages it was able to draw on its broad product range to offer more attractive alternatives to its faithful customers. At the same time, its global production empire meant that it was able to supplement changing local demand in the US with exports from Japan to the degree that output at home actually rose. As far as anyone could tell at the time, it was situation normal with the Japanese showing their flexibility by rapidly exploiting the market shift away from heavy sport utility vehicles (SUVs). Presumably, if the markets had then stabilized, the Japanese manufacturers would have readjusted their grip on the United States to a stranglehold. Then, as 2008 staggered to a close, the Japanese edifice too began to crumble.

In assuming that future volume growth would come in the emerging markets, the industry had forgotten that the most lucrative value added sales are in the developed world. Furthermore, one market, the United States, is so stuffed full of acquisitive, wealthy consumers that all the cost advantages of mass production can be fully realized and handsome profits earned. Flexible manufacturing, using single plants to produce a range of products, may be emblematic of the most sophisticated manufacturers, but it is only by dedicating plants to the making of single, standard products that all the economies of scale can be fully exploited. When factories were operating at full speed, churning out Toyota Camrys and Honda Accords, for example, each loaded up with expensive optional extras, the manufacturers might as well have been printing money. It was not just the big names either, all the Japanese manufacturers had looked to the United States for the majority of their profits. This golden era ended in 2008.

If only the Japanese had properly globalized. Although companies like to claim that globalization is part of their corporate structure, true integration of a firm's international operations is rare. Ford and GM have progressed most of the way, having entirely separate North American and European operations. Even so, it is only now that they are calling on their European divisions to provide small-car technology, once considered unsuited to American tastes, that the level of international integration that can be called globalization is being attained.

The Japanese, meanwhile, are two steps behind in the globalization process. They have always maintained their core operations at home, only extending facilities into overseas locations as the occasion warranted. This was a brilliant strategy while the Japanese currency was undervalued, allowing overseas plants to meet the bulk of local demand while Japanese plants exported more specialist high value vehicles. But the sharp reduction of mass market demand in the United States, and the rise in the value of the Japanese yen, has left this strategy in tatters. All the Japanese manufacturers, with their reliance on exports, are feeling the pain, but Toyota most of all has shocked with the stunning reversal in its fortunes and the threat of deepening operational losses.

This strategic failure fully to embrace the United States as a second home for Toyota is what will put it at a severe disadvantage over the long-term future. While Ford and GM learnt to join the ranks of local manufacturers in the European scene, Toyota has only managed to exploit weakness in the United States. As the local Detroit Big 3 concentrated on making profits on light truck-based SUVs, which could be made like commercial vehicles but sold as luxury cars, the Japanese were given almost free rein in the passenger car market. Both sides were happy as long as the American mass market was split between light trucks and cars.

The flaw in the Japanese approach, though, was that they did this with neat, fuel-efficient cars designed in the European idiom rather than the chrome-laden land-yachts of the Big 3. With the local brands offering little credible opposition in the car sector it was difficult to judge whether American fashion had changed, shifted to the SUV behemoths or simply gone into hibernation.

The sudden popularity of the new Chrysler 300C passenger car at its launch in 2005, boasting all the styling grace of a custom battle tank, indicated that the Americans had not yet lost their taste for the tasteless. While the 300C could never have threatened Japanese hegemony by itself, it was ominous that the Japanese had no response within their range of more effete car models. In any case, the Japanese were busy opening a new front into the Big 3's light truck sector with their own monstrous creations. If the oil price had stayed low, and the US economy remained on cruise control, then this attack on the American's last redoubt could have finished off Detroit for good. In the nick of time, the credit crunch and the recession intervened to hammer the light truck sector and force the Big 3 to confront their impending doom.

The age of the SUV and the light truck sector is not yet dead, but it is clear that future growth in the United States market will be in the passenger car market. Since cars are appreciably lighter than SUVs they offer the kind of fuel consumption that is better suited to rising oil prices and tighter emission regulations. However, only Detroit is grabbing the spotlight with full-size cars that demonstrate the return of assertiveness in chrome and steel that elsewhere in the world would be called brash. Leading Japanese hybrid models such as the Toyota Prius and the Honda Insight may seem sensible offerings to European eyes, but next to the latest Detroit stars they come over as pious and even a little dull. Worse still, in the short term low fuel prices and tight money create classic conditions for the kind of technically simple and relatively low cost models in which Detroit has always excelled.

Until at least 2010, the future of the domestic American automotive industry hangs in the balance, but Detroit is defending its last redoubt with the courage of the condemned. Thanks to government backing, they should emerge with a solid foundation for future growth. This will put the squeeze on the Japanese, who will find they are losing presence in a market that they had previously used as their private bank. The Japanese will not be repelled from the market, they have built up too much presence for that, but the white heat of local competition and the high value of the yen will render the market much less profitable.

In lieu of a divine automotive being this might be seen as economic justice. Japanese manufacturers have enjoyed a charmed period with a low value yen and a submissive American car market. Now that these factors have turned against them, they have a choice between battening down the hatches and surviving the economic storm as best as they can, or to come out fighting. Such a proactive a strategy would emulate the historic infiltration of Ford and GM into Europe. But Ford did this when the infant European industry was barely out of the nursery, not an epithet that would describe the fight for survival now taking place in the United States. GM took the swifter route into Europe of buying the established local brands,

Opel of Germany and Vauxhall of the UK. Such a bold maneuver in the United States would provide instant access to the very heart of the American industry. Through radical consolidation a Japanese manufacturer would provide common vehicle platforms for a vast range of global model variants, whilst simultaneously serving the peculiarities of the US market with locally designed, authentic Americana. In North America today there is only one integrated manufacturer up for sale, and only one Japanese manufacturer with the resources to take it on: Chrysler and Toyota.

Sadly, it looks like Toyota is paralyzed by the fear of the abyss it is staring into. After decades of cutting production costs to the bone, its announcements that it will find salvation in unearthing even further savings lack credibility. To put it simply, if the opportunity to cut costs had existed, Toyota would have already acted. Honda is busy making its own cutbacks, losing various halo models which were once considered so crucial but now summarily discarded. Nissan, too, is wielding the axe while Mitsubishi looks to be sinking into a death spiral. Only Subaru, with its tightly grouped range of distinctive vehicles, held US sales steady during 2008 but it too has suffered financially as the currency exchange rates inverted.

We will have the answer by the end of 2010. If the economic earthquake reduces the Big 3 to rubble then the Japanese will emerge to take over the US market largely unopposed. But if the Big 3 survive, and chances are that at least two of them will, then they will be retaking all the major segments of their home market with self-confident, American-designed automobiles. This will push the Japanese manufacturers into accepting less lucrative shares of other markets around the world, and the diminished profit margins will mean a drastic pruning of their extensive model development programs. As the Japanese are the leaders in fuel cell research we may have to accept that the long-cherished dream of a release from the influence of oil will be postponed yet again. Whatever the outcome, it seems that the automobile paradigm will be with us for a while yet.

Bibliography

ACEA (2007) *Automotive Production (EU + EFTA Countries) Summary 2006*, accessed July 16, 2007, at www.acea.be/files/PRODUCTION%202006.pdf.

Adeney, M. (1988) *The Motor Makers: The Turbulent History of Britain's Car Industry*. London: Collins.

Almor, T., Hashai, N., and Hirsch, S. (2006) The product cycle revisited: Knowledge intensity and firm internationalization, *Management International Review*, vol. 46, no. 5, pp. 507–626.

Altman, M. (1999) The methodology of economics and the survival principle revisited and revised: some welfare and public policy implications of modeling the economic agent, *Review of Social Economy*, vol. 57, no. 4, pp. 427–49.

Altshuler, A., Anderson, M., Jones, D., Roos, D., and Womak, J. (1986) *The Future of the Automobile: The Report of MIT's International Automobile Program*. Cambridge, MA: MIT Press.

Álvarez Gil, M.J., and González de la Fé, P. (1999) Strategic alliances, organisational learning and new product development: the cases of Rover and Seat, *R&D Management*, vol. 29, no. 4, pp. 423–6.

Amicus (2004) *Motor Vehicles Fact Sheet*. London: Amicus.

Anastakis, D. (2004) From independence to integration: the corporate evolution of the Ford Motor Company of Canada, 1904–2004, *Business History Review*, 78, pp. 213–53.

Andersen, O. (1993) On the internationalization process of firms: a critical analysis, *Journal of International Business Studies*, 24 (2), p. 209.

Ankarloo, D., and Palermo, G. (2004) Anti-Williamson: a Marxian critique of New Institutional Economics, *Cambridge Journal of Economics*, 28, pp. 413–29.

Antràs, P. (2005) Incomplete contracts and the product cycle, *American Economic Review*, vol. 95, no. 4, pp. 1054–73.

Armour, H.O., and Teece, D.J. (1978) Organizational structure and economic performance: a test of the multidivisional hypothesis, *Bell Journal of Economics*, vol. 9, no. 1, pp. 106–22.

Armour, H.O., and Teece, D.J. (1980) Vertical integration and technological innovation, *Review of Economics and Statistics*, vol. 62, no. 3, pp. 470–4.

Artz, K.W., and Brush, T.H. (2000) Asset specificity, uncertainty and relational norms: an examination of coordination costs in collaborative strategic alliances, *Journal of Economic Behaviour and Organization*, vol. 41, pp. 337–62.

Assembly (2002) Then and now: the mechanical marvel, *February 1*, accessed March, 2002, at www.assemblymag.com/CDA/ArticleInformation/features/BNP__Features__Item/0,6493,98331,00.html.

Autocar (2001) New car prices, December 12.

Autocar (2005) How £100 m could save MG, April 19, pp. 46–9.

Automotive Design and Production (2004) Nissan: how to achieve shift, from website, July, accessed September 14, 2006, at www.autofieldguide.com/articles/070401.html.

Automotive Engineer (2004a) In the right mind, January, pp. 60–1.

Automotive Engineer (2004b) Rover seeks Proton alliance in survival battle, March, pp. 4–5.

Automotive News (2000) New Rover Cars owner ponders possible partners, May 15, p. 42.

Automotive News (2001) MG Rover brings suppliers to Longbridge, January 29, p. 40.

Automotive News (2004a) *2004 Market Data Book*, supplement, June 28.

Automotive News (2004b) Mitsubishi crisis is round 2 for Gilligan, August 2, p. 24.

Automotive News (2004c) *Guide to Global Automotive Partnerships*, supplement, October 4.

Automotive News (2005a) *2005 Market Data Book*, supplement, May 23.

Automotive News (2005b) Asian's new R&D Taj Mahals are a sign of success, May 16, p. 1.

Automotive News (2005c) Subaru plan finally has one mission, June 13, p. 24.

Automotive News (2005d) *Guide to Global Automotive Partnerships*, supplement, August.

Automotive News (2006a) *2006 Market Data Book*, supplement, May 22.

Automotive News (2006b) *2006 Global Market Data Book*, supplement, June 26.

Automotive News (2006c) *2006 Guide to China's Auto Market*, supplement, May 1.

Automotive News (2006d) Magna Steyr: big changes needed to be competitive, August 14.

Automotive News (2007a) *2007 Global Market Data Book*, supplement, June 25.

Automotive News (2007b) *2007 Guide to China's Auto Market*, supplement, May 14.

Automotive News (2007c) Hyundai halts production at Alabama plant, October 5.

Automotive News (2007d) Even sold, Aston works with Ford, March 19.

Automotive News (2008) *2008 Market Data Book*, supplement, June 23.

Automotive News Europe (2006) *2006 Market Data Book*, supplement, May.

AutoTech Daily (2002) GM touts flexible body shop tooling, December 16, p. 1.

Autoweek (1999) What Morgan for the millennium?, November 1, accessed August 23, 2006, at www.gomog.com/articles/autoweek.html.

Aw, B.-Y. and Batra, G. (1998) Firm size and the pattern of diversification, *International Journal of Industrial Organization*, vol. 16, pp. 313–31.

AWKnowledge (2003) Capacity utilization data, August/September.

Bailey, D. (2003) Globalisation, regions and cluster policies: the case of the Rover Task Force, *Policy Studies*, vol. 24, nos. 2/3, pp. 67–85.

Bailey, D. (2007) Globalization and restructuring in the auto industry: the impact on the West Midlands automobile cluster, *Strategic Change*, vol. 16, pp. 137–44.

Bailey, D., Sugden, R., and Harte, G. (1994) *Making Transnationals Accountable: A Significant Step for Britain*. London: Routledge.

Bardgett, L. (2000) *Rover: The Story So Far,* Parliamentary Research Paper 00/41, http://www.parliament.uk/commons/lib/research/rp2000/rp00-041.pdf.

Barkema, H.G., Bell, J.H.J., and Pennings, J.M. (1996) Foreign entry, cultural barriers and learning, *Strategic Management Journal*, vol. 17(2), pp. 151–66.

Barnet, R.J., and Müller, R.E. (1974) *Global Reach: The Power of the Multinational Corporations*. New York: Simon & Schuster.

Bartlett, C.A., and Ghoshal, S. (1993) Beyond the M-Form: toward a managerial theory of the firm, *Strategic Management Journal*, vol. 14, pp. 23–46.

Beamish, P.W., and Makino, S. (1998) Performance and survival of joint ventures with non-conventional ownership structures, *Journal of International Business Studies*, vol. 29, pp. 797–818.

Bell, J., McNaughton, R., and Young, S. (2001) 'Born again' global firms: an extension to the 'born global' phenomenon, *Journal of International Management*, vol. 7, no. 3, pp. 1–17.

Benady, D. TGWU (2003) Jobs relocation: a one-way street?, TGWU publication, December 1, accessed April 5, 2005, at www.tgwu.org.uk/Templates/Journal.asp?NodeID=90352.

Benson, J.K. (1983) A dialectical method for the study of organisations, in Morgan, G. (ed.) *Beyond Method: Strategies for Social Research*. Beverly Hills, CA: Sage.

Berger, S. (2000) Globalization and politics, *Annual Review of Political Science*, vol. 3, pp. 43–62.

Beynon, H. (1984) *Working for Ford*. Harmondsworth: Pelican.

Bhaskar (1979) *The Future of the UK Motor Industry*. London: Kogan Page.

Blois, K.J. (1972) Vertical quasi-integration, *Journal of Industrial Economics*, vol. 20, no. 3, pp. 253–72.

BMW (2007) *Company Facts – Sustainability and the Environment*, accessed October 13, 2007, at www.bmweducation.co.uk/default.asp.

Borensztein, E., De Gregorio, J., and Lee, J-W (1998) How does foreign direct investment affect economic growth?, *Journal of International Economics*, vol. 45, pp. 115–35.

Brady, C., and Lorenz, A. (2001) *End of the Road: BMW and Rover: A Brand too Far*. London: Pearson Education.

British Psychological Society (1992) Ethical principles for conducting research with human participants from British Psychological Society, accessed November 1, 2006, at www.bps.org.uk/the-society/ethics-rules-charter-code-of-conduct/code-of-conduct/ethical-principles-for-conducting-research-with-human-participants.cfm.

Bryman, A. (2001) *Social Research Methods*. Oxford: Oxford University Press.

Bryman, R. (1989) *Research Methods and Organization Studies*. London: Routledge.

Buckley, M., and Rees, C. (2002) *The World Encyclopaedia of Cars: The Definitive Guide to Classic and Contemporary Cars from 1945 to the Present Day*. London: Hermes House.

Buckley, P.J. (1990) Problems and developments in the core theory of international business, *Journal of International Business Studies*, vol. 21, Iss. 4, pp. 657–65.

Buckley, P.J. (2002) Is the international research agenda running out of steam?, *Journal of International Business Studies*, vol. 33, no. 2, pp. 365–73.

Buckley, P.J., and Casson, M. (1976) *The Future of the Multinational Enterprise*. London: Macmillan.

Buckley, P.J., and Casson, M. (1981) The optimal timing of a foreign direct investment, *Economic Journal*, vol. 91, no. 361, pp. 75–87.

Buckley, P.J., and Casson, M. (1998a) Models of the multinational enterprise, *Journal of International Business Studies*, vol. 29, no. 1, pp. 21–44.

Buckley, P.J., and Casson, M.C. (1998b) Analyzing foreign market entry strategies: extending the internalization approach, *Journal of International Business Studies*, vol. 29, no. 3, pp. 539–61.

Buckley, P.J., and Ghauri, P.N. (2004) Globalisation, economic geography and the strategy of multinational enterprises, *Journal of International Business Studies*, vol. 35, pp. 81–98.

Burton, R.B., and Obel, B. (1980) A computer simulation test of the M-Form hypothesis, *Administrative Science Quarterly*, vol. 25, no. 3, pp. 457–66.

Business Week (2000) Unloading Rover may not win the race, April 3, p. 59.

Business Week (2007) China's auto industry: roaring ahead, May 2.

Cabral, L.M.B. (2000) *Introduction to Industrial Organization.* Cambridge, MA: MIT Press.

Cairncross, A. (1966) *Introduction to Economics.* London: Butterworth.

Calvet, A.L. (1981) A synthesis of foreign direct investment theories and theories of the multinational firm, *Journal of International Business Studies*, vol. 12, pp. 43–59.

Cantner, U., Dreßler, K., and Kruger, J.J. (2006) Firm survival in the German automobile industry, *Empirica*, vol. 33, pp. 49–60.

Carr, C. (1993) Global, national and resource-based strategies: an examination of strategic choice and performance in the vehicle components industry, *Strategic Management Journal*, vol. 14, no. 7, pp. 551–67.

Carr, C. (1999) Globalisation, strategic alliances, acquisitions and technology transfer. Lessons from ICL/Fujitsu and Rover/Honda and BMW, *R&D Management*, vol. 29, Iss. 4, pp. 405–21.

Caves, R.E. (1996) *Multinational Enterprise and Economic Analysis.* Cambridge: Cambridge University Press.

Central Policy Review Staff (1975) *The Future of the British Car Industry: A Report.* London: Her Majesty's Stationery Office.

Chalmers, A.F. (1983) *What is this Thing Called Science?* Milton Keynes: Open University Press.

Chandler, A.D. (1977) *The Visible Hand: The Managerial Revolution in American Business.* London: Harvard University Press.

Chandler, A.D. (1990) *Strategy and Structure: Chapters in the History of the American Industrial Enterprise.* Cambridge, MA: MIT Press.

Chang, S.J., and Choi, U. (1988) Strategy, structure and performance of Korean business groups: a transactions cost approach, *Journal of Industrial Economics*, vol. 37, no. 2, pp. 141–58.

Chase, J. (1995) Historical analysis in psychological research, in Breakwell G.M., Hammond S. and Fife-Schaw C. (eds) *Research Methods in Psychology.* London: Sage.

Cheng, L.K., and Kwan, Y.K. (2000) What are the determinants of the location of foreign direct investment? The Chinese experience, *Journal of International Economics*, vol. 51, pp. 379–400.

Chery (2007) Chery Engine Company, accessed August 9, 2007, at www.cheryacteco.com/en/index.jsp.

Church, R. (1979) *Herbert Austin: The British Motor Car Industry to 1941*. London: Europa.

Church, R. (1994) *The Rise and Decline of the British Motor Industry*. London: Macmillan.

Clark, K.B., Chew, W.B.T., Fujimoto, Meyer J., and Scherer, F.M. (1987) Product development in the world auto industry, *Brookings Papers on Economic Activity*, vol. 1987, no. 3, pp. 729–81.

Clarke, R. (1987) Conglomerate firms, in Clarke, R., and Mc Guiness, A. (eds) *The Economics of the Firm*. Oxford: Basil Blackwell.

Cleeve, E. (1997) The motives for joint ventures: a transaction cost analysis of Japanese MNEs in the UK, *Scottish Journal of Political Economy*, vol. 44, no. 1, pp. 31–43.

Coase, R.H. (2000) The acquisition of Fisher Body by General Motors, *Journal of Law and Economics*, 43, pp. 15–31.

Coles, J.W., and Hesterly, W.S. (1998) Transaction costs, quality, and economies of scale: examining contracting choices in the hospital industry, *Journal of Corporate Finance*, 4, pp. 321–45.

Collins, P., and Stratton, M. (1993) British Car Factories from 1896: *A Complete Historical, Geographical, Architectural and Technological Survey*. Godmanstone, Dorset: Veloce.

Cooper, A.C. (1964) R&D is more efficient in small companies, *Harvard Business Review*, 42 (3), pp. 75–83.

Corbetta, P. (tr. Bernard Patrick) (2003) *Social Research: Theory, Methods and Techniques*. London: Sage.

Cowling, K., and Sugden, R. (1998) Strategic trade policy reconsidered: national rivalry vs free trade vs international cooperation, *Kyklos*, vol. 51, Iss. 3, pp. 339–57.

Cowling, K., and Tomlinson, P.R. (2005) Globalisation and corporate power, *Contributions to Political Economy*, vol. 24, pp. 33–54.

Creswell, J.W. (2003) *Research Design: Qualitative, Quantitative and Mixed Methods Approaches*. London: Sage.

Cusumano, M. (1985) *The Japanese Automobile Industry: Technology and Management at Nissan and Toyota*, Cambridge, MA: Harvard University Press.

DaimlerChrysler (2004) An engineering odyssey: from all-steel to true unit bodies 1914–1960 from Walter P Chrysler Museum, accessed April 7, 2005, at www.chryslerheritage.com/sec500_pdf/An%20Engineering%20Odyssey.pdf.

Dankbaar, B. (1992) *Economic Crisis and Institutional Change: The Crisis of Fordism from the Perspective of the Automobile Industry.* Maastricht: UPM.

D'Aveni, R.A., and Ravenscraft, D.J. (1994) Economies of integration versus bureaucracy costs: does vertical integration improve performance?, *Academy of Management Journal*, vol. 37, no. 5, pp. 1167–206.

David, R.J., and Han, S-K (2004) A systematic assessment of the empirical support for transaction cost economics, *Strategic Management Journal*, 25, pp. 39–58.

Demizu, T. (2003) *Honda: Its Technology and Management.* Osaka: Union Press.

Denscombe, M. (1998) *The Good Research Guide: For Small-scale Social Research Projects.* Buckingham: Open University Press.

Denzin, N.K. (1978) *The Research Act: A Theoretical Introduction to Sociological Methods* (2nd edn). New York: McGraw-Hill.

De Vaus, D. (2001) *Research Design in Social Research.* London: Sage.

Dexter, L.A. (1970) *Elite and Specialised Interviewing.* Evanston, IL: Northwestern University Press.

Dey, I. (1993) *Qualitative Data Analysis: A User-friendly Guide for Social Scientists.* London: Routledge.

Dicken, P. (2003) *Global Shift: Transforming the World Economy.* London: Sage.

Diesel Progress (1999) Peugeot Citroen and Ford sign diesel engine agreement, December, accessed August 15, 2007, at http://findarticles.com/p/articles/mi_m0FZX/is_12_65/ai_58674994.

Donaldson, L. (2001) *The Contingency Theory of Organizations.* London: Sage.

Donnelly, T., Morris, D., and Donnelly, T. (2005) Renault-Nissan: a marriage of necessity?, *European Business Review*, vol. 17, Iss. 5, pp. 428–40.

Doz, Y.L., and Prahalad, C.K. (1991) Managing DMNCs: a search for a new paradigm, *Strategic Management Journal*, vol. 12, pp. 145–64.

Drucker, P.F. (1972) *Concept of the Corporation.* New York: John Day.

DTI (2003) *The 2003 R&D Scoreboard.* London: HMSO.

DTI (2004) *The 2004 R&D Scoreboard.* London: HMSO.

DTI (2005) *The 2005 R&D Scoreboard.* London: HMSO.

DTI (2006) *The 2006 R&D Scoreboard.* London: HMSO.

DTI/Auto Industry (2007) Background information on MG Rover, accessed February 8, 2007, at www.autoindustry.co.uk/features/MG%20Rover/MG_Rover_Background.

Dunne, P., and Hughes, A. (1994) Age, size, growth and survival: UK companies in the 1980s, *Journal of Industrial Economics*, vol. 42, no. 2, pp. 115–40.

Dunnett, P.J.S. (1980) *The Decline of the British Motor Industry: The Effects of Government Policy, 1945–1979*. London: Croom Helm.

Dunning, J.H. (1979) Explaining changing patterns of international production: in defence of the eclectic theory, *Oxford Bulletin of Economics and Statistics*, vol. 41, Iss. 4, pp. 269–95.

Dunning, J.H. (1980) Toward an eclectic theory of international production: some empirical tests, *Journal of International Business Studies*, vol. 11, pp. 9–31.

Dunning, J.H. (1981) *International Production and the Multinational Enterprise*. London: George Allen & Unwin.

Dunning, J.H. (1995) Reappraising the eclectic paradigm in an age of alliance capitalism, *Journal of International Business Studies*, vol. 26, Iss. 3, pp. 461–91.

Dunning, J.H. (2000) The eclectic paradigm as an envelope for economic and business theories of MNE activity, *International Business Review*, 9, pp. 163–90.

Dussauge, P., Garrette, B., and Mitchell, W. (2004) Asymmetric performance: the market share impact of scale and link alliances in the global auto industry, *Strategic Management Journal*, vol. 25, Iss. 7, pp. 701–11.

Dyer, J.H. (1996) Does governance matter? Keiretsu alliances and asset specificity as sources of Japanese competitive advantage, *Organization Science*, vol. 7, no. 6, pp. 649–66.

Dyer, J.H. (1997) Effective interfirm collaboration: how firs minimize transaction costs and maximise transaction value, *Strategic Management Journal*, vol. 18, Iss. 7, pp. 535–56.

Economist (2000) Walking away from Longbridge, March 16.

Economist (2002) Special Report: Car Manufacturing, February 23.

Economist (2005) The global car industry: extinction of the predator, September 8.

Edwardes, M. (1983) *Back from the Brink: An Apocalyptic Experience*. London: Collins.

Erdener, C., and Shapiro, D.M. (2005) The internationalisation of Chinese family enterprises and Dunning's eclectic MNE paradigm, *Management and Organization Review*, vol. 1, no. 3, pp. 411–36.

Etemad, H., Wright, R.W., and Dana, L.P. (2001) Symbiotic international business networks: collaboration between small and large firms, *Thunderbird International Business Review*, vol. 43 (4), pp. 481–99.

Etter, J.-F. and Perneger, T.V. (2000) Snowball sampling by mail: application to a survey of smokers in the general population, *International Journal of Epidemiology*, 29, pp. 43–8.

Faugier, J., and Sargeant, M. (1997) Sampling hard to reach populations, *Journal of Advanced Nursing*, 26 (4), pp. 790–7.

Ferguson, P.R., and Ferguson, G.J. (2000) *Organisations: A Strategic Perspective*. Basingstoke: Macmillan.

Feyerabend, P. (1993) *Against Method*. London: Verso.

Fife-Schaw, C. (1995) Surveys and sampling issues, in Breakwell, G.M., Hammond, S., and Fife-Schaw, C. (eds) *Research Methods in Psychology*. London: Sage.

Financial Times (2001) How Jensen got back on the road, August 23, p. 17.

Financial Times (2002a) A driving ambition to do the impossible, May 3, p. 14.

Financial Times (2002b) Defiant car maker sees a positive route to the future, April 27, p. 3.

Financial Times (2005) MG Rover plans to recruit 200 engineers and development staff, March 1, p. 2.

Fitz, J., and Halpin, D. (1995) Brief encounters: researching education policy-making in elite settings, in Salisbury, J., and Delamont, S. (eds) *Qualitative Studies in Education*. Aldershot: Ashgate.

Flick, U. (2004) Triangulation in qualitative research, in Flick, U., von Kardoff, E., and Steinke, I. (eds) Jenner, B. *A Companion to Qualitative Research*. London: Sage.

Ford (2004) Welcome to the home of the Ford Motor Company and our family of brands, accessed November 6, 2006, at www.ford.com/en/default.htm?source=topnav.

Ford (2006) Ford flexible engine production boosted by investment in Britain, Ford press release 06/098, August 22.

Freeland, R.F. (1996) The myth of the M-Form? Governance, consent, and organizational change, *American Journal of Sociology*, vol. 102, no. 2, September, pp. 483–526.

Frech, H.E., and Ginsberg, P.B. (1974) Optimal scale in medical practice: a survivor analysis, *Journal of Business*, vol. 47, Iss. 1 (January), pp. 23–36.

Frech, H.E., and Mobley, L.R. (1995) Resolving the impasse on hospital scale economies: a new approach, *Applied Economics*, vol. 27, no. 3, pp. 286–96.

Freeland, R.F. (1996a) The myth of the M-form: governance, consent and organizatiuonal change, *American Journal of Sociology*, vol. 102, no. 2 (September), pp. 483–526.

Freeland, R.F. (1996b) The Struggle for Control of the Modern Corporation: Organizational Change at General Motors, 1924–1958, *Business and Economic History*, vol. 25, no. 1, pp. 32–7.

Freeland, R.F. (2000) Creating holdup through vertical integration: Fisher Body revisited, *Journal of Law and Economics*, vol. 43, pp. 33–66.

Fuel Cells 2000 (2007) Online Fuel Cell Information Resource, accessed October 13, 2007, at www.fuelcells.org/.

Fumia Design (2006) Lancia J: tribute to Lancia century, Fumia Design press release, October, accessed January 8, 2007, at www.fumiadesign.com/sel_news.html.

Garrahan, P., and Stewart, P. (1991) Nothing new about Nissan?, in Law, C.M. (ed.) *Restructuring the Global Automobile Industry: National and Regional Impacts*. London: Routledge.

Gear Product News (2006) Cast gear blanks: what to consider when, June, pp. 34–9.

Georgano, N. (2000) *The Beaulieu Encyclopaedia of the Automobile*. London: The Stationery Office.

Gershwin, S.B. (1987) An efficient decomposition method for the approximate evaluation of tandem queues with finite storage space and blocking, *Operations Research*, vol. 35, no. 2, pp. 291–305.

Gilson, R.J., and Roe, M.J. (1993) Understanding the Japanese keiretsu: overlaps between corporate governance and industrial organization, *Yale Law Journal,* vol. 102, no. 4, pp. 871–906.

Gorg, H., and Strobl, E. (2003) 'Footloose' multinationals, *Manchester School*, vol. 71, no. 1, pp. 1–19.

Grossman, S.J., and Hart, O.D. (1986) The costs and benefits of ownership: a theory of vertical and lateral integration, *Journal of Political Economy*, vol. 94, no. 4, pp. 691–719.

Hagel, J., and Singer, M. (1999) Unbundling the corporation, *Harvard Business Review*, pp. 133–41.

Hakim, C. (1987) *Research Design*. London: Allen & Unwin.

Hamel, J. with Dufour, S., and Fortin, D. (1991) *Case Study Methods*. London: Sage.

Harrigan, K.R. (1986) Matching vertical integration strategies to competitive situations, *Strategic Management Journal*, vol. 7, pp. 535–55.

Harland & Wolff (2007) Engineering Design Resources, accessed October 17, 2007, at www.harland-wolff.com/facilities1.asp.

Hart, O., and Moore, J. (1990) Property rights and the nature of the firm, *Journal of Political Economy*, vol. 98, no. 6, pp. 1119–58.

Harvey-Jones, J. with Masey, A. (1990) *Trouble Shooter.* London: BBC Books.

Hedlund, G. (1994) A model of knowledge management and the N-form corporation, *Strategic Management Journal*, vol. 15, pp. 73–90.

Hennart, J-F and Reddy, S. (1997) The choice between mergers/acquisitions and joint ventures: the case of Japanese investors in the United States, *Strategic Management Journal*, vol. 18, pp. 1–12.

Hill, C.W.L. (1985) Internal organization and enterprise performance: some UK evidence, *Managerial and Decision Economics*, vol. 6, Iss. 4, pp. 210–16.

Hill, C.W.L. (1988) Internal capital market controls and financial performance in multidivisional firms, *Journal of Industrial Economics*, vol. 37, no. 1, pp. 67–83.

Hill, C.W.L. (2003) *International Business: Competing in the Global Marketplace.* Boston: McGraw-Hill.

Hill, C.W.L., Hwang, P., and Kim, W.C. (1990) An eclectic theory of the choice of international entry mode, *Strategic Management Journal,* vol. 11, pp. 117–28.

Ho, Y.C., Eyler, M.A., and Chien, T.T. (1983) A new approach to determine parameter sensitivities to transfer lines, *Management Science*, vol. 29, no. 6, pp. 700–14.

Holmstrom, B., and Milgrom, P. (1994) The firm as an incentive system, *The American Economic Review*, vol. 84, no. 4, pp. 972–91.

Holweg, M., and Oliver, N. (2005) Who killed MG Rover?, special report from the Cambridge-MIT Institute's Centre for Competitiveness and Innovation, April 25.

Honda (1999) *A Dynamic Past, An Exciting Future.* Tokyo: Honda.

Honda (2002) *2002 FactBook.* Tokyo: Honda.

Honda (2004) Honda's US automobile operations, *Honda Annual Report 2004*, p. 15.

Honda (2006) *2006 FactBook.* Tokyo: Honda.

Horaguchi, H., and Toyne, B. (1990) Setting the record straight: Hymer, internationalization theory and transaction cost economics, *Journal of International Business Studies*, vol. 21, Iss. 3, pp. 487–94.

Hoskisson, R.E., Hill, C.W.L., and Kim, H. (1993) The multidivisional structure: organizational fossil or source of value?, *Journal of Management*, vol. 19, no. 2, pp. 269–98.

Hounshell, D.A. (1984) *From the American System to Mass Production, 1800–1932.* London: Johns Hopkins University Press.

Humphries, A.S., and Wilding, R.D. (2004) Long term collaborative business relationships: the impact of trust and C^3 behaviour, *Journal of Marketing Management,* vol. 20, pp. 1107–22.

Husan, R. (1997) The continuing importance of economies of scale in the automotive industry, *European Business Review*, vol. 97, no. 1, pp. 38–42.

Hymer, S.H. (1960) The international operations of national firms, a study of direct foreign investment, PhD thesis, Massachusetts Institute of Technology.

Hymer, S. (1970) The efficiency (contradictions) of multinational corporations, *American Economic Review*, vol. 60, no. 2, pp. 441–8.

Hymer, S. (1972) The internationalization of capital, *Journal of Economic Issues*, vol. 6, Iss. 1, pp. 91–111.

Hymer, S., and Pashigian, P. (1962) Turnover of firms as a measure of market behaviour, *Review of Economics and Statistics*, vol. 44, no. 1, pp. 82–7.

Iran Khodro (2005) Paykan history, accessed June 20, 2005, at www.ikco.com/products/paykan.asp.

JAMA (2005) Growing in America, accessed August 30, 2006, at www.jama.org/library/factsheets/Contributions_2005.pdf.

JAMA (2006) *2006: The Motor Industry of Japan*. Tokyo: Japan Automobile Manufacturers Association.

JAMA (2007a) *2007: The Motor Industry of Japan*. Tokyo: Japan Automobile Manufacturers Association.

JAMA (2007b) Japan's motor vehicle statistics, accessed September 24, 2007, at www.jama.org/statistics/index.htm.

Janesick, V.J. (2000) The choreography of qualitative research design, in Denzin, N.K., and Lincoln, Y.S. (eds) *Handbook of Qualitative Research* (2nd edn). London: Sage.

Japan Times (2008) Toyota unit signs contract to build new unit near Sendai, February 22, accessed March 23, at www.japantimes.co.jp.

Jick, T.D. (1979) Mixing qualitative and quantitative methods: triangulation in action, *Administrative Science Quarterly*, vol. 24, Iss. 4, pp. 602–11.

Johanson, J., and Vahlne, J.-E. (1990) The mechanism of internationalisation, *International Marketing Review*, vol. 7, Iss. 4, pp. 11–24.

Jones, J., and Wren, C. (2006) *Foreign Direct Investment and the Regional Economy*. Aldershot: Ashgate.

Joskow, P.L. (1985) Vertical integration and long-term contracts: the case of coal-burning electric generating plants, *Journal of Law, Economics and Organization,* vol. 1 (1), pp. 33–80.

Joskow, P.L. (2003) Vertical integration, in Ménard, C., and Shirley, M.M. (eds) *Handbook of New Institutional Economics.* Cambridge, MA: Springer.

Just-Auto (2004) General Motors to relocate Asia-Pacific HQ to Shanghai and expand local design centre, June 23, accessed September 20, 2006, at http://just-auto.com/news_detail.asp?art=44815&dm=yes.

Kakabadse, A., and Kakabadse, N. (2000) Sourcing: new face to economies of scale and the emergence of new organizational forms, *Knowledge and Process Management*, vol. 7, no. 2, pp. 107–18.

Kia (2007) Kia history, accessed August 9, 2007, at www.kiainfo.co.uk/about.htm#Kia%20History.

Kindleberger, C.P. (1984) *Multinational Excursions.* Cambridge, MA: MIT Press.

Kitay, J., and Callus, R. (1998) The role and challenge of case study design in industrial relations research, in Whitfield, K. and Strauss, G. (eds) *Researching the World of Work: Strategies and Methods in Studying Industrial Relations.* Ithaca, NY: Cornell University Press.

Klein, B. (1980) Transaction cost determinants of "unfair" contractual arrangements, *American Economic Review*, 70, pp. 356–62.

Klein, B. (1988) Vertical integration as organizational ownership: the Fisher Body–General Motors relationship revisited, *Journal of Law, Economics, and Organization*, vol. 4, no. 1, Spring, pp. 199–213.

Klein, B. (2000) Fisher-General Motors and the nature of the firm, *Journal of Law and Economics*, 43, pp. 105–41.

Klein, S., Frazier, G.L., and Roth, V.J. (1990) A transaction cost analysis model of channel integration in international markets, *Journal of Marketing Research*, vol. 27, pp. 196–208.

Klepper, S. (1996) Entry, exit, growth, and innovation over the product life cycle, *The American Economic Review*, vol. 86, no. 3, pp. 562–83.

Klepper, S. (2002a) Firm survival and the evolution of oligopoly, *RAND Journal of Economics*, vol. 33, no. 1, pp. 37–61.

Klepper, S. (2002b) The capabilities of new firms and the evolution of the U.S. automobile industry, *Industrial and Corporate Change*, vol. 11, no. 4, pp. 645–66.

Kogut (1991) Joint ventures and the option to expand and acquire, *Management Science*, vol. 37, no. 1, pp. 19–33.

Kolodny, H.F. (1979) Evolution to a matrix organisation, *Academy of Management Review*, vol. 4, no. 4, pp. 543–53.

Koutsoyiannis, A. (1979) *Modern Microeconomics* (2nd edn). London: Macmillan.

KPMG (2007) Momentum 2007 KPMG Global Auto Executive Survey, accessed July 17, 2007, at www.kpmg.co.uk/pubs/ACFFCBC.pdf.

Krafcik, J.F. (1988) Triumph of the lean production system, *Sloan Management Review*, vol. 30 (1), pp. 41–52.
Kuemmerle, W. (1999) Foreign direct investment in industrial research in the pharmaceutical and electronics industries: results from a survey of multinational firms, *Research Policy*, vol. 28, pp. 179–93.
Kuhn, A.J. (1986) *GM Passes Ford, 1918–1938*. London: Pennsylvania State University Press.
Kuhn, T.S. (1970) *The Structure of Scientific Revolutions*. London: University of Chicago Press.
Lakatos, I. (1970) Methodology of scientific research programmes, in Lakatos, I., and Musgrave, A. (eds) *Criticism and the Growth of Knowledge*. Cambridge: Cambridge University Press.
Lamming, R. (1993) *Beyond Partnership: Strategies for Innovation and Lean Supply*. London: Prentice Hall.
Lee, T.W. (1999) *Using Qualitative Methods in Organizational Research*. London: Sage.
Leung, M.K., Young, T., and Digby, D. (2003) Explaining the profitability of foreign banks in Shanghai, *Managerial and Decision Economics*, vol. 24, pp. 15–24.
Levy, D. (1985) Transactions cost approach to vertical integration: an empirical examination, *Review of Economics and Statistics*, vol. 67, no. 3, pp. 438–45.
Lewchuck, W. (1987) *American Technology and the British Vehicle Industry: A Century of Production in Britain*. Cambridge: Cambridge University Press.
Lierberman, M.B. (1991) Determinants of vertical integration: an empirical test, *Journal of Industrial Economics*, vol. 39, no. 5, pp. 451–66.
Lord, R.G., and Hohenfeld, J.A. (1979) Longitudinal field assessment of equity effects on the performance of major league baseball players, *Journal of Applied Psychology*, vol. 64, Iss. 1, pp. 19–26.
Lotus (2006) Lotus Engineering – The Total Solution Provider, accessed September 1, 2006, at www.grouplotus.com/generic/generic.php?section=17&page=21&page_id=21.
Lyons, B.R. (1995) Specific investment, economies of scale, and the make-or-buy decision: a test of transaction cost theory, *Journal of Economic Behaviour and Organization*, vol. 26, pp. 431–43.
Mair, A. (1994) *Honda's Global Local Corporation*. Basingstoke: Macmillan.
Makadok (1999) Interfirm differences in scale economies and the evolution of market shares, *Strategic Management Journal*, 20, pp. 935–52.
Malmberg, A. (1991) Restructuring the Swedish manufacturing industry: the case of the motor vehicle industry, in Law, C.M. (ed.) *Restructuring*

the Global Automobile Industry: National and Regional Impacts. London: Routledge.

Mariti, P., and Smiley, R.H. (1983) Co-operative agreements and the organization of industry, *Journal of Industrial Economics*, vol. 31, Iss. 4, pp. 437–51.

Masten, S.E., Meehan, J.W., and Snyder, E.A. (1991) The costs of organization, *Journal of Law, Economics and Organization*, vol. 7, no. 1, pp. 1–25.

Maxcy, G. (1981) *The Multinational Motor Industry*. London: Croom Helm.

Maxcy, G., and Silberston, A. (1959) *The Motor Industry*. London: Allen & Unwin.

Maxton, G.P., and Wormald, J. (1995) *Driving Over a Cliff? Business Lessons from the World's Car Industry*. Wokingham: EIU/ Addison-Wesley.

Mazda (2004) Mazda History, accessed November 6, 2006, at http://media.ford.com/mazda/article_display.cfm?article_id=15147.

McAfee, R.P., and McMillan, J. (1995) Organizational diseconomies of scale, *Journal of Economics and Management Strategy*, vol. 4, no. 3, pp. 399–426.

Mehrabi, M.G., Ulsoy, A.G., and Koren, Y. (2000) Reconfigurable manufacturing systems: key to future manufacturing, *Journal of Intelligent Manufacturing*, vol. 11(4), pp. 403–19.

Merriam, S.B. (1988) *Case Study Research in Education: A Qualitative Approach*. London: Jossey-Bass.

Miles, M.B. (1979) Qualitative data as an attractive nuisance: the problem of analysis, *Administrative Science Quarterly*, vol. 24, Iss. 4, pp. 590–601.

Miles, M.B., and Huberman, A.M. (1994) *Qualitative Data Analysis: An Expanded Sourcebook* (2nd edn). London: Sage.

Milgrom, P., and Roberts, J. (1992) *Economics, Organisation and Management*. Englewood Cliffs, NJ: Prentice Hall.

Miller, K.D., and Reuer, J.J. (1998) Firm strategy and economic exposure to foreign exchange rate movements, *Journal of International Business Studies*, vol. 29, pp. 493–513.

MIRU (Motor Industry Research Unit) (1988) *Rover: Profile, Progress and Prospects. An Appraisal and Forecast of the Rover Group*. Norwich: Motor Industry Research Unit.

Monteverde, K., and Teece, D.J. (1982) Supplier switching costs and vertical integration in the automobile industry, *Bell Journal of Economics*, vol. 13, no. 1, pp. 206–13.

Morris, J., and Imrie, R. (1992) *Transforming Buyer–Supplier Relations*. London: Macmillan.

Moschandreas, M. (1994) *Business Economics*. London: Routledge.

Moyser and Wagstaff (eds)(1987) *Research Methods for Elite Studies*. London: Allen & Unwin.

Munck, R. (2002) *Globalisation and Labour: The New "Great Transformations."* London: Zed Books.

Nash, T. (2002) GM's futuristic "skateboard" in *Motor*, April, 2002, accessed October 15, 2007, at http://findarticles.com/p/articles/mi_qa3828/is_200204?pnum=4&opg=n9055816.

Nieuwenhuis, P., and Wells, P. (1997) *The Death of Motoring?: Car Making and Automobility in the 21st Century*. Chichester: Wiley.

Nieuwenhuis, P., and Wells, P. (2002) *The Automotive Industry: The Future Guide*. London: CAIR-BT.

Nieuwenhuis, P., and Wells, P. (2003) *The Automotive Industry and the Environment*. Cambridge: Woodhead Publishing.

Nieuwenhuis, P., and Wells, P. (2007) The all-steel body as a cornerstone to the foundations of the mass production car industry, *Industrial and Corporate Change*, vol. 16, Iss. 2, pp. 1–29.

Observer (2004) What's Rover driving at?, April 4, 2004, p. 3.

O'Callaghan, T.J.O. (2002) *The Aviation Legacy of Henry and Edsel Ford*. Ann Arbor, MA: Proctor Publications.

Odaka, K., Ono, K., and Adachi, F. (1988) *The Automobile Industry in Japan: A Study of Ancillary Firm Development*. Tokyo: Kinokuniya.

OICA (2006a) *The World's Automotive Industry: Some Key Figures*, November 29, 2006, accessed July 16, 2007, at www.oica.net/htdocs/Main.htm.

OICA (2006b) Production statistics 1998–2005, accessed November 5, 2006, at www.oica.net/htdocs/Main.htm.

OICA (2008a) Economic contributions, accessed September 19, 2008, at http://oica.net/.

OICA (2008b) *Production statistics 1998–2007*, accessed 13 December 2008 at www.oica.net/category/production-statistics.

Panzar, J.C., and Willig, R.D. (1981) Economies of scope, *American Economic Review*, vol. 71, no. 2, pp. 268–72.

Parkin, M. (1996) *Economics*. Wokingham: Addison-Wesley.

Parkin, M., Powell, M., and Matthews, K. (1997) *Economics* (3rd edn). Harlow: Addison-Wesley.

Pascale, R.T., and Athos, A.G. (1986) *The Art of Japanese Management*. Harmondsworth: Penguin.

Payne, S.L. (1951) *The Art of Asking Questions*. Princeton, NJ: Princeton University Press.

Pilkington, A. (1999) Strategic alliance and dependency in design and manufacture: the Rover-Honda case, *International Journal of Operations and Production Management*, vol. 19, Iss. 5/6, pp. 460–73.

Pires, S.R.I. (1998) Managerial implications of the modular consortium model in a Brailian automotive plant, *International Journal of Operations and Production Management*, vol. 18, Iss. 3, pp. 221–32.

Plutchik, R. (1983) *Foundations of Experimental Research*. London: Harper & Row.

Polanyi, K. (1963) *The Great Transformation: The Political and Economic Origins of our Time*. Boston, MA: Beacon Press.

Poppo, L., and Zenger, T. (1998) Testing alternative theories of the firm: transaction cost, knowledge-based, and measurement explanations for make-or-buy decisions in information services, *Strategic Management* Journal, vol. 19, pp. 853–77.

Porter, M.E. (1986) Competition in global industries: a conceptual framework, in Porter, M.E. (ed.) *Competition in Global Industries*. Boston MA: Harvard Business School Press.

Pratten, C.F. (1971) *Economies of Scale in the Manufacturing Industry*. Cambridge: Cambridge University Press.

PricewaterhouseCoopers (2005) Global Automotive Financial Review. New York: PricewaterhouseCoopers.

Professional Engineering (2003) Rover's bumpy road to recovery, July 23, pp. 26–7.

Punch, K.F. (2005) *Introduction to Social Research: Quantitative and Qualitative Approaches* (2nd edn). London: Sage.

Rae, J.B. (1984) *The American Automobile Industry*. Boston, MA: Twayne Publishers.

Rayome, D., and Baker, J.C. (1995) Foreign direct investment: a review and analysis of the literature, *International Trade Journal*, vol. 9, no. 1, pp. 3–37.

Reliable Plant (2007) 2007 Harbour Report: Toyota leads in total productivity, July–August 2007, accessed August 1, 2007, at www.reliableplant.com/article.asp?articleid=6572.

Rhys, G. (1972) *Motor Industry: An Economic Survey*. London: Butterworths.

Rhys, G. (1999) Japan's approach to globalization, *Business Briefing: Global Automotive Manufacturing and Technology*, January.

Ricketts, M. (1987) *The Economics of Business Enterprise: New Approaches to the Firm*. Brighton: Wheatsheaf Books.

Robson, G. (1988) *The Rover Story*. Cambridge: Stephens.

Rogers, R.P. (1993) The minimum optimal steel plant and the survivor technique of cost estimation, *Atlantic Economic Journal*, vol. 21, no. 3, pp. 30–7.

Rosegger, G. (1986) *The Economics of Production and Innovation*. Oxford: Pergamon Press.
Rossman, G.B., and Rallis, S.F. (1998) *Learning in the Field: An Introduction to Qualitative Research*. London: Sage.
Rugman, A., and Hodgetts, R. (2001) The end of global strategy, *European Management Journal*, vol. 19, No 4, pp. 333–43.
Ryen, A. (2004) Ethical issues, in Seal, C., Gobo, G., Gubrium, J.F., and Silverman, D. (eds) *Qualitative Research Practice*. London: Sage.
Sako, M. (2006) *Shifting Boundaries of the Firm*. Oxford: Oxford University Press.
Scarbrough, H., and Terry, M. (1996) *Industrial Relations and Reorganization of Production in the UK Motor Industry: A Study of the Rover Group*. Coventry: Industrial Relations Research Unit.
Schott, P.K. (2003) One size fits all? Heckscher-Ohlin specialization in global production, *American Economic Review*, vol. 93, no. 3, June, pp. 686–708.
Sharratt, B. (2000) *The Men and Motors of "The Austin": The Inside Story of a Century of Car Making at Longbridge*. Yeovil: Haynes Publishing.
Shepherd, W.G. (1967) What Does the Survivor Technique Show about Economies of Scale?, *Southern Economic Journal*, vol. 34, no. 1, pp. 113–22.
Shimokawa, K. (2003) Production innovation at Ford of Europe's Cologne Plant. *Nikkan Jidosha Shumbun*, December 13, 2003, accessed August 24, 2006, at http://imvp.mit.edu/papers/0304/shimokawa_cologne.pdf.
SID (2006) R&D Centres: Tata Motors Ltd, Society for Innovation & Development, Indian Institute of Science Campus, Bangalore, accessed August 24, 2006, at http://sid.iisc.ernet.in/tml.html.
Sieradski, L. (2002) Opel Corsa: the accidental world car, accessed November 27, 2006, at http://globaledge.msu.edu/NewsAndViews/views/papers/opel_corsa_accidental_world_car.pdf.
Silberston, A. (1972) Economies of scale in theory and practice, *Economic Journal*, vol. 82, no. 325, pp. 369–91.
Silver, B. (2003) *Forces of Labour: Workers' Movements since 1870*. Cambridge: Cambridge University Press.
Slater, G., and Spencer, D.A. (2000) The uncertain foundations of transaction cost economics, *Journal of Economic Issues*, vol. 34, no. 1, pp. 61–87.
Sloan, A.P. (1986) *My Years with General Motors*. London: Penguin.
Sloman, J. (2001) *Essentials of Economics*. Harlow: Pearson Education.
Smith, D. (1991) Industrial restructuring and the labour force: the case of Austin Rover in Longbridge, Birmingham, in Law, C.MN. (ed.) *Restructuring the Global Automobile Industry: National and Regional Impacts*. London: Routledge.

Smitka, M.J. (1991) *Competitive Ties: Subcontracting in the Japanese Automotive Industry*. New York: Columbia University Press.

SMMT (2006a) *Motor Industry Facts – 2005*. London: SMMT.

SMMT (2006b) Manufacturing plants profile: Toyota Motor Manufacturing (UK) Ltd, SMMT press release, August.

Stadco (2006) Stadco signs contract with Nanjing Automobile Corporation (UK), press release, October 16, accessed March 12, 2007, at www.stadco.co.uk/news/2006-10-16.

Stake, R.E. (1995) *The Art of Case Study Research*. London: Sage.

Stake, R.E. (1998) Case studies, in Denzin, N.K., and Lincoln, Y.S. (eds) *Strategies of Qualitative Inquiry*. London: Sage.

Stake, R.E. (2000) Case study research methods, University of Illinois, unpublished lecture notes, accessed April 13, 2002, at www.ed.uiuc.edu/CIRCE/EDPSY399/.

Stigler, G.J. (1958) Economies of scale, *Journal of Law and Economics*, vol. 1, pp. 54–71.

Sturgeon, T.J. (2000) How do we define value chains and production networks?, background paper prepared for Bellagio Value Chains Workshop, September 25–October 1, Bellagio, Italy.

Sturgeon, T., and Florida, R. (2000) Globalization and jobs in the automotive industry, final report to the Alfred P Sloan Foundation, New York.

Sunday Times (2006) BAE to get Nimrod lift-off, July 16, p. 3.

Teece, D.J. (1980) The diffusion of administrative innovation, *Management Science*, vol. 26, no. 5, pp. 464–70.

Teece, D.J. (1981) Internal organization and economic performance: an empirical analysis of the profitability of principal firms, *Journal of Industrial Economics*, vol. 30, no. 2, pp. 173–99.

Thomas, R.J. (1995) Interviewing important people in big companies, in Hertz, R., and Imber, J.B. (eds) *Studying Elites Using Qualitative Methods*. London: Sage.

Thompson, A.A., and Formby, J.P. (1993) *Economics of the Firm: Theory and Practice*. London: Prentice Hall International.

Thoms, D., and Donnelly, T. (1985) *The Motor Car Industry in Coventry since the 1890s*. London: Croom Helm.

Times (2007) Chinese deal further blow to Longbridge, accessed February 10, 2007, at http://business.timesonline.co.uk/tol/business/industry_sectors/engineering/article1303182.ece.

Times 100 (2004) Nissan: planning for quality and productivity, accessed September 18, 2006, at www.thetimes100.co.uk/downloads/nissan/nissan_9_full.pdf.

Tirole, J. (1989) *The Theory of Industrial Organization*. Cambridge, MA: MIT Press.

Tolliday, S. (1998) The diffusion and transformation of Fordism: Britain and Japan compared, in Boyer, R., Charron, E., Jurgens, U., and Tolliday, S. (eds) *Between Imitation and Innovation: The Transfer and Hybridization of Productive Models in the International Automobile Industry*. Oxford: Oxford University Press.

Toyota (2007a) Global Vision 2010, accessed September 25, 2007, at www.toyotatr.com/eng/gvision2010.asp.

Toyota (2007b) Toyota in the world 2007, accessed September 25, 2007, at www.toyota.co.jp/en/about_toyota/in_the_world/pdf2007/databook_2007.pdf.

Truett, L.J., and Truett, D.B. (2001) The Spanish automotive industry: scale economies and input relationships, *Applied Economics*, vol. 33, no. 12, pp. 1503–13.

Truett, L.J., and Truett, D.B. (2003) The Italian automotive industry and economies of scale, *Contemporary Economic Policy*, vol. 21, no. 3, pp. 329–37.

UBS/Ricardo (2007) Is diesel set to boom in the US?, accessed October 15, 2007, at www.ricardo.com/download/pdf/R119361S.pdf.

UNCTAD (2002) *World Investment Report, 2002*. New York: United Nations.

Underwood, J. (1989) *Will to Win: John Egan and Jaguar*. London: W.H. Allen.

Vernon, R. (1966) International investment and international trade in the product cycle, *Quarterly Journal of Economics*, vol. 80, no. 2, pp. 190–207.

Vernon, R. (1979) The product cycle hypothesis in a new international environment, *Oxford Bulletin of Economics and Statistics*, vol. 41, no. 4, pp. 255–67.

Viitanen, H. (2002) Internationalization process and theories, Swedish School of Economics and Business Administration, accessed January 18, 2007, at www.shh.fi/~polsa/inteory/.

VM Motori, (2006) Current vehicles, accessed September 7, 2006, at www.vmmotori.it/en/01/01/index.jsp.

Volkswagen (2005) Annual report 2005, accessed March 23, 2007, at www.volkswagen.co.uk/assets/pdf/AnnualReport2005.pdf.

Walker, G., and Weber, D. (1987) Supplier competition, uncertainty, and make-or-buy decisions, *Academy of Management Journal*, vol. 30, no. 3, pp. 589–95.

Ward's Automotive Year Books. Detroit: Ward's Inc.

Weijian, S. (1991) Environmental risks and joint venture sharing arrangements, *Journal of International Business Studies*, vol. 22, Iss. 4, pp. 555–78.

Wells, P., and Nieuwenhuis, P. (2001) *The Automotive Industry: A Guide*. London: CAIR-BT.

Wells, P., and Rawlinson, M. (1994) *The New European Automobile Industry*. Basingstoke: Macmillan.

Whipp, R., and Clark, P. (1986) *Innovation and the Auto Industry: Product, Process and Work Organization*. London: Frances Pinter.

Wilding, R., and Humphries, A.S. (2006) Understanding collaborative supply chain relationships through the application of the Williamson organisational failure framework, *International Journal of Physical Distribution and Logistics Management*, vol. 36, no. 4, pp. 309–29.

Wilkins, M. (1979) *The Maturing of Multinational Enterprise: American Business Abroad from 1914 to 1970*. Cambridge, MA: Harvard University Press.

Wilks, S. (1990) Institutional insularity: government and the British motor industry since 1945, in Chick, M. (ed.) *Governments, Industries and Markets: Aspects of Government–Industry Relations in the UK, Japan, Germany and the USA since 1945*. Aldershot: Elgar.

Williams, B. (1997) Positive theories of multinational banking: eclectic theory versus internalisation theory, *Journal of Economic Surveys*, vol. 11, Iss. 1, pp. 71–100.

Williams, K., Haslam, C., Johal, S., and Williams, J. (1994) *Cars: Analysis, History, Cases*. Providence, RI: Berghahn Books.

Williams, K., Williams, J., and Haslam, C. (1987) *The Breakdown of Austin Rover: A Case Study in the Failure of Business Strategy and Industrial Policy*. Leamington Spa: Berg.

Williamson, O.E. (1973) Markets and hierarchies: some elementary considerations, *American Economic Review*, vol. 63, no. 2, pp. 316–25.

Williamson, O.E. (1975) *Markets and Hierarchies: Analysis and Antitrust Implications. A Study in the Economics of Internal Organization*. London: Collier Macmillan.

Williamson, O.E. (1981) The economics of organization: the transaction cost approach, *American Journal of Sociology*, 87, 3, pp. 548–77.

Williamson, O.E. (1983) Credible commitments: using hostages to support exchange, *American Economic Review*, vol. 73, no. 4, pp. 519–40.

Williamson, O.E. (1987) *Antitrust Economics: Mergers, Contracting and Strategic Behaviour*. Oxford: Basil Blackwell.

Williamson, O.E. (1991) Comparative economic organization: the analysis of discrete structural alternatives, *Administrative Science Quarterly*, vol. 36, pp. 269–96.

Williamson, O.E. (1993) Opportunism and its critics, *Managerial and Decision Economics*, vol. 14 , pp. 97–107.

Williamson, O.E. (1996) Economic organization: the case for candor, *Academy of Management Review*, vol. 21, no. 1. 48–57.

Williamson, O.E. (1998) Transaction cost economics: how it works; where it is headed, *De Economist*, vol. 146, no. 1, pp. 23–58.

Williamson, S., and Wright, R. (1994) Barter and monetary exchange under private information, *American Economic Review*, vol. 84, no. 1, pp. 104–23.

Womack, J.P., Jones, D.T., Ross, D. (1990) *The Machine that Changed the World*. New York: Rawson Associates/Macmillan.

Wood, J. (1988) *Wheels of Misfortune: The Rise and Fall of the British Motor Industry*. London: Sidgwick & Jackson.

World Business Council for Sustainable Development (WBCSD)(2004) Mobility 2030: meeting the challenges to sustainability, accessed October 15, 2007, at www.wbcsd.org/web/publications/mobility/overview.pdf.

World Economic Forum (2002) Mahathir Bin Mohamad, Malaysian Prime Minister, Speech to The World Economic Forum – East Asia Economic Summit 2002, accessed June 15, 2003, at www.smpke.jpm.my/website/webdb.nsf/?Opendatabase.

Yamaguchi, J.K. (2003) *The Mazda RX-8: The World's First 4-door, 4-seat Sports Car*. Kanagawa: Ring.

Yamin, M. (1991) A reassessment of Hymer's contribution to the theory of transnational corporations, in Pitelis, C., and Sugden, R. (eds) *The Nature of the Transnational Firm*. London: Routledge.

Yin, R. (1994) *Case Study Research: Design and Methods*. London: Sage.

Yin, R. (2006) Case study methods, in Green, J.L., Camilli, G., and Elmore, P.B. (eds) *Handbook of Applied Social Research Methods*. London: Sage.

Youn Kim, H. (1986) Economies of scale and economies of scope in multiproduct financial institutions: further evidence from credit unions, *Journal of Money, Credit, and Banking*, vol. 18, no. 2, pp. 220–6.

Zenger, T.R. (1994) Explaining organization diseconomies of scale in R&D: agency problems and the allocation of engineering talent, ideas, and effort by firm size, *Management Science*, vol. 40, no. 6, pp. 708–29.

INDEX